FITNESS MEDICINE

Edited by **Hasan Sözen**

Fitness Medicine
http://dx.doi.org/10.5772/62558
Edited by Hasan Sözen

Contributors

Yati Boutcher, Jeanne Grace, Daniel Gaz, Arnulfo Ramos-Jiménez, Rebeca Chavez-Herrera, Aida S. Castro-Sosa, Rosa P Hernández-Torres, David Olivas-Davila, Iván Chulvi-Medrano, Juan Tortosa, Juan Manuel Cortell-Tormo, José Sanchis, Selma Cırrık, Gülay Hacıoğlu, Hidetaka Hamasaki

Published by InTech

Janeza Trdine 9, 51000 Rijeka, Croatia

Publishing Process Manager Iva Simcic
Technical Editor SPi Global
Cover InTech Design team

First published October, 2016
Printed in Croatia

Additional hard copies can be obtained from orders@intechopen.com

Fitness Medicine, Edited by Hasan Sözen
p. cm.
Print ISBN 978-953-51-2744-4
Online ISBN 978-953-51-2745-1

For EU product safety concerns:
IN TECH d.o.o., Prolaz Marije Krucifikse Kozulić 3, 51000 Rijeka, Croatia,
info@intechopen.com or visit our website at intechopen.com.

Meet the editor

Dr. Hasan Sözen is an assistant professor in the School of Physical Education and Sport at Ordu University (Ordu, Turkey). Her primary research interest includes sport and exercise physiology. Dr. Sözen received his PhD degree in Health Science Institute Department of Physical Education and Sport at Ondokuz Mayıs University (Samsun, Turkey). His postdoctoral fellowship in the field of electromechanical delay in the Department of Biomedical Sciences for Health at the University of Milan (Milan, Italy) was supported by the Scientific and Technological Research Council of Turkey.

Contents

Preface

Although fitness and health have similar properties, they are, in reality, two very different concepts. While health refers to the absence of diseases, fitness refers to the degree of body functioning and the ability of the body to handle physical demands. The more efficient the body functions, the higher the level of fitness. The higher the level of fitness, the greater the chance of the body being free of diseases and maintaining a healthy state. Fitness is a major part of a preventative medicine approach to health. Utilization of fitness is supported by researchers to have more healthy more than one hundred years. To be fit not only refers to be healthy but also refers to be protected from diseases. This book, we believe, will cover and enlighten those aspects where fitness has begun to play important roles nowadays.

The book begins with an overview of exercise for medicine by Boutcher, where the role of exercise as preventative medicine for a disease-free lifestyle. After the first chapter, there comes a chapter about body shape, imagination, and composition through fitness performance by Ramos-Jimenez et al. After the second chapter, there comes an interesting chapter on activity-tracking devices by Gaz et al. The next chapter by Cirrik and Hacioglu summarizes a very important, yet mostly underestimated, subject: neurophysiological effects of exercise. The chapter serves its aim and provides a global understanding of neurophysiological effects of exercise parameters used to identify cognition, learning, memory, depression, aging, and neurodegenerative diseases. The fifth chapter by Chulvi et al. introduces a novel analysis about exercise for hypertension. What are the effects of resistance training on autonomic nervous function in older individuals? This striking question's answer is investigated in a research by Hamasaki. His chapter provides details on resistance training, sarcopenia, autonomic nervous function, and heart rate variability evidenced in his well-written findings. The last chapter which is written by Grace describes an interesting task performed with acquired immune deficiency syndrome (AIDS). Her chapter is about exercise therapy for acquired immune deficiency syndrome patients. I believe that this chapter will provide a new perspective for the patients.

This book may of course have errors despite our obsessive reviews and efforts. But all in all, I think that it provides the reader with interesting up-to-date data while summarizing some basic fitness medicine knowledge.

As a last remark, I want to thank all the authors of this book for their amazing works and our publishing process manager Ms. Iva Simcic, without whom I would not be able to edit this book.

I dedicate this book to all the people from whom I have learned a lot of things in this life: to my teachers, my colleagues, my subjects, my students, and mostly to my beloved son Kuzey Sözen. I know I could not manage life and my career without him.

I wish and hope that this book will be useful for anyone who wants to read about new perspectives in fitness medicine. I also hope that it will arouse a new and great inspiration for researchers working in this field.

Assist. Prof. Dr. Hasan Sözen
Ordu University, School of Physical Education and Sport
Ordu, Turkey

Chapter 1

Exercise is Medicine: The Importance of Exercise as Preventative Medicine for a Disease-Free Lifestyle

Yati N. Boutcher

Additional information is available at the end of the chapter

http://dx.doi.org/10.5772/64981

Abstract

Currently, overweightness and obesity are the biggest health problem faced in the twenty-first century. The major causes of this problem are lack of physical activity and excessive consumption of processed food. Individuals who are overweight typically show abnormal cardiovascular function, and obesity is an independent risk factor for cardiovascular diseases, such as hypertension, heart disease, and type 2 diabetes. Hypertension has been found to occur more frequently in overweight compared to lean individuals and poses a risk for hypertension development in young overweight adults. Obesity has also been found to be associated with reduced life expectancy and sudden death largely through its negative effect on the cardiovascular system. Degradation of a number of key autonomic cardiovascular markers, such as reduced heart rate variability and baroreflex sensitivity, is associated with the development of these lifestyle diseases. Also alterations in the metabolic profile, such as dyslipidemia and hyperglycemia, can lead to impairment in cardiac structure and function. Regular exercise has been widely used as preventative medicine to reverse autonomic, cardiovascular, and metabolic decline. Thus, incorporating regular exercise into daily activity may prevent the development of these cardiovascular diseases and accompanying risk factors.

Keywords: exercise, cardiovascular health, metabolic health, arterial stiffness, heart rate variability, baroreflex sensitivity

1. Introduction

Physical inactivity has been shown to affect health by increasing cardiovascular and metabolic dysfunction. More and more people including children and adolescents are leading a sedentary lifestyle, which leads to a number of health problems. Overweightness and obesity are often accompanied by a lack of physical activity that can lead to more serious health problems, such as metabolic syndrome, type 2 diabetes, and hypertension. Lack of physical activity has also been found to be a major cause of morbidity [1] and mortality [2]. Overweight and obese individuals typically show abnormal cardiovascular function, and obesity has been found to be an independent risk factor for cardiovascular diseases, such as hypertension [3], type 2 diabetes, and cardiac complications (i.e., heart failure, heart disease). Obesity is associated with reduced life expectancy [4] and sudden death through its negative effect on the cardiovascular system [5]. Overweight and obese individuals usually develop hypertension. Compared to normal weight individuals, overweightness and obesity pose a potential risk for hypertension development in overweight individuals. It was found that a 10 kg excess in weight was associated with a 3 mmHg higher systolic and a 2–3 mmHg higher resting diastolic blood pressure. This can be translated into an estimated increased risk for coronary heart disease of 12% and an increased risk of stroke of 24% [6]. Alterations in the metabolic profile that are commonly found in overweight and obese individuals, such as dyslipidemia and hyperglycemia, can also lead to impairment in cardiac structure and function as adipose tissue accumulates [7]. It has been shown that left ventricular (LV) function and the ratio of the stroke index and LV end-diastolic pressure are reduced with overweightness and obesity [8]. These changes show that depressed LV function has already occurred even in young overweight and obese individuals.

Overwhelming evidence indicates that regular physical activity in the form of both acute and regular aerobic exercise reduces the severity and occurrence of diseases related to unhealthy lifestyles. One single bout of acute exercise has been found to improve vascular function even in normotensive young healthy individuals with a family history of hypertension [9]. Whereas regular aerobic exercise has been shown to improve physical fitness (cardiorespiratory fitness) in addition to improving vascular function [10]. Having high levels of physical fitness, assessed through a sub-maximal or maximal oxygen uptake test, is desirable as low cardiorespiratory fitness has been found to significantly increase the risk of cardiovascular diseases and mortality more so than other factors, such as hypertension, type 2 diabetes, and smoking, and regardless of body mass index [11]. High-intensity interval training (HIIT) is a form of interval sprinting exercise typically performed on a stationary bike and has also been found to improve cardiac and metabolic health of young overweight males [12] and females [13]. Physical activity is so important that failure to lead a physically active lifestyle can result in several abnormalities such as high blood pressure, metabolic syndrome, and type 2 diabetes [14]. Baroreceptor sensitivity (BRS) has also been found to decrease with age [15, 16], being hypertensive [17] and being overweight and obese [18].

Overall, regular exercise results in improvement of the pathogenesis and symptoms of specific conditions that include chronic heart failure, coronary heart disease, dyslipidemia, hypertension, insulin resistance, intermittent claudication, obesity, and type 2 diabetes [19]. Other health benefits of regular aerobic exercise include improvement in balance, cognitive functioning, life expectancy, and overall quality of life [20, 21]. The beneficial effects of regular aerobic exercise have been well known. Thus, modification of a sedentary lifestyle by incorporating regular exercise is paramount for maintaining and improving health. Exercise is a powerful stimulus and can reduce the severity of several conditions with its effects being similar to many drug therapies. In this regard, exercise has now been widely accepted as medicine [22]. Therefore, the purpose of the chapter is to summarize the major autonomic, cardiovascular, and metabolic changes associated with being physically inactive. Then, the ability of differing forms of exercise (aerobic, interval sprinting, and resistance) to reverse these negative changes will be described.

2. Cardiovascular problems associated with physical inactivity

A sedentary lifestyle, which is associated with overweightness and obesity, typically leads to an increased risk of cardiovascular disease. Currently, overweightness and obesity have been escalating at an alarming rate worldwide. Based on the World Health Organization (WHO, Global Health Observatory data) region data from 2010 to 2014 [23], the prevalence of overweightness (**Figure 1**) and obesity (**Figure 2**) for both males and females at the age of ≥18 years has significantly increased. The highest prevalence of overweightness and obesity, according to the WHO, was the Americas, followed by Europe and South Pacific, with Southeast Asia containing the lowest prevalence of overweightness and obesity. Also, women were more likely to be obese than men in 2014 in all WHO regions. Increased numbers of overweight and obese individuals put tremendous pressure and burden on healthcare providers and threatens world health in many countries. Approximately 20–30% of adults worldwide are categorized as clinically obese and numbers are progressively increasing [24, 25]. It has been well established that both high body mass index and a sedentary lifestyle are associated with greater risk of cardiovascular disease [26, 27].

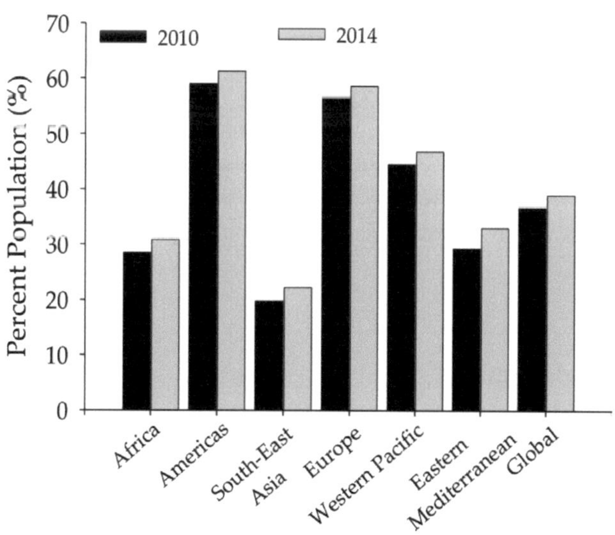

Figure 1. Prevalence of overweight based on WHO region for both sexes 2010–2014.

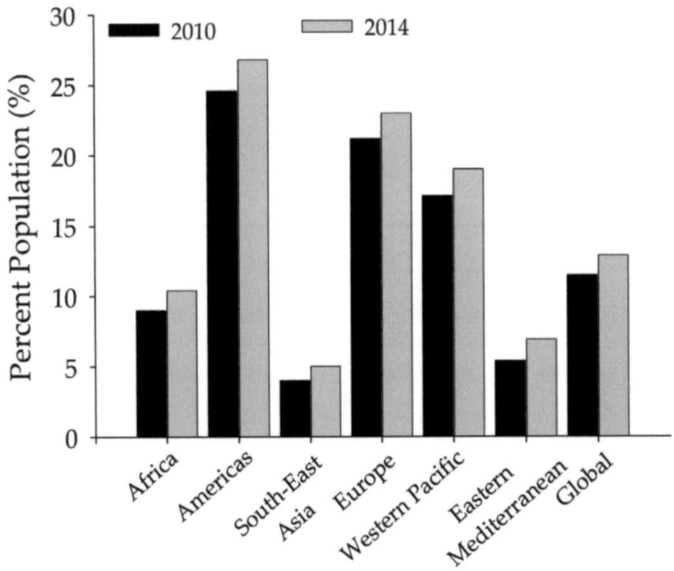

Figure 2. Prevalence of obesity based on the WHO region for both sexes 2010–2014.

It has been shown that overweightness and obesity are accompanied by a cluster of cardiovascular risk factors, which is termed metabolic syndrome and includes hypertension, glucose intolerance, hypertriglyceridemia, and visceral obesity. These conditions occur in approximately one out of four adults over the age of 40 [28]. The metabolic syndrome is also associated with insulin resistance and endothelial dysfunction [29, 30]. It is believed that in child obesity, as fat mass increases, insulin resistance develops, which is a determinant of impaired metabolic function at early age [31]. Other health problems accompanying overweightness and obesity include cardiac autonomic dysfunction and endothelial dysfunction. Aberrant cardiac autonomic function has been found in obese children [32] and increased adiposity during childhood increases the risk of obesity, type 2 diabetes, and cardiovascular disease in adulthood [33, 34]. Endothelial dysfunction can be defined as inadequate endothelial-mediated vasodilation, which is typically due to a deficiency of endothelial-derived relaxing factor, nitric oxide (NO) synthesis, and/or release. It is believed that NO deficiency is a primary factor linking insulin resistance and endothelial dysfunction. Thus, it is clear that being sedentary results in increased cardiovascular disease risk, aberrant metabolic, cardiac, autonomic, and endothelial function.

2.1. Cardiac autonomic function

It has been well established that cardiac autonomic dysfunction is associated with conditions, such as overweight, obesity, and type 2 diabetes [35, 36]. The decline of cardiac autonomic function is typically influenced by an accumulation of visceral fat [37]. Lowered autonomic function was also found to be correlated with higher abdominal-to-peripheral body fat distribution measured by dual energy X-ray absorptiometry in both young and old healthy

http://dx.doi.org/10.5772/64981

men [38]. These results suggest that visceral obesity contributes to a decline in autonomic function. Also, young individuals with high abdominal adiposity seemed to undergo an early decline and a premature aging of autonomic function [37]. Thus, hypertension and high levels of central adiposity have an unfavorable effect on cardiac autonomic function, which is reflected by impaired heart rate variability (HRV) and baroreflex sensitivity (BRS).

HRV is a marker that is commonly used to assess cardiac autonomic function and can be defined as beat-to-beat variation in the heart rate of individuals possessing normal sinus rhythm [39]. HRV has been shown to be an indicant of healthy cardiac autonomic function as it has been found that reduced HRV predicts increased cardiovascular disease and mortality [40, 41]. High body mass index, which is commonly found in overweight and obese individuals, is also associated with reduced HRV, which contributes to decreased parasympathetic activity and increased sympathetic activity [42, 43]. BRS is another marker of cardiac autonomic function that provides information about the ability to increase parasympathetic or vagal activity and to decrease sympathetic activity in response to sudden increases in blood pressure [44]. BRS, which is typically reduced as people age, is an indicant of the body's autonomic nervous system sensitivity in responding to sudden blood pressure changes [44]. Low BRS is a marker of reduced compliance of the carotid artery [15, 16]. Overweight and obese individuals typically possess endothelial dysfunction that leads to reduced arterial compliance, which is accompanied with a low BRS compared to that of normal weight individuals. Lower BRS has also been found to be associated with high blood pressure, increased sympathetic activity, and diseases related to an unhealthy lifestyle. It seems that low HRV and BRS are affected by the accumulation of visceral fat and metabolic syndrome factors.

Studies have shown that obese children typically have decreased parasympathetic activity compared to normal weight children [45–49]. It has also been reported that an increased thickness of carotid intima-media in obese children represents a very early sign of atherosclerosis [50]. It has also been found that the earliest signs of atherosclerosis development are lipid deposits in the intimal layer of systemic arteries (aorta), which have been found in children as early as 3 years old and in the coronary arteries of adolescents [51]. Another aberrant change possessed by overweight and obese individuals is high levels of leptin. Several studies have shown that positive correlations among leptin, insulin resistance, and blood pressure existed in overweight children [52, 53]. A reduction in adiposity level, however, brought about a reduction in leptin, which further improved metabolic health, by decreasing insulin, cardiovascular function, and blood pressure [52]. Adiponectin has a major anti-inflammatory and anti-atherogenic effect [54] and has been found to be three times higher in normal weight children compared to overweight and obese children [34]. A 1-year follow-up, following a 1-year intervention, found that children with weight loss had similar adiponectin levels compared to normal weight children. Leptin levels, however, remained higher in overweight and obese children with weight loss. The increase in adiponectin appears to be an early biomarker of improvement in insulin sensitivity [34]. Thus, it is clear that overweightness and obesity in children result in early deposition of fat in the systemic arteries and thickening of arterial walls. These conditions can lead to more serious health complications later in their life if lifestyle modification is not introduced.

2.2. Vascular dysfunction

Overweight and obesity conditions not only result in metabolic dysfunction but also bring about vascular dysfunction. The balance between the ability of the vasculature (arterioles) to vasodilate and vasoconstrict is impaired with the development of overweightness and obesity. The loss of balance between vasodilators and vasoconstrictors is an indicant of endothelial dysfunction. Endothelial dysfunction is now regarded as a precursor of atherogenesis [55], which further leads to the development of atherosclerosis [56] and diabetic complications [57]. Arterial compliance, and its inverse arterial stiffness, can be defined as the ability of an artery to distend in response to intravascular (transmural) pressure [58] and is commonly termed as stiffening of the arteries. It is well known that overweight and obese individuals typically have high arterial stiffness [59], and this high arterial stiffness is associated with high body mass index and body fat [60]. Reduction in body mass index, however, has been found to be the strongest determinant of decreased arterial stiffness in severe obese young and middle-aged adults [61].

Arterial stiffness, which has been considered to be an independent predictor of cardiovascular disease, can be assessed through a reflection wave, commonly known as the augmentation index (AIx) and carotid femoral pulse wave velocity (PWVcf), which is considered to be the gold standard of arterial stiffness assessment. AIx, a reflection wave, is typically assessed by placing an applanation tonometry sensor (e.g., SphygmoCor) on either a radial or carotid artery. AIx then is derived from the ratio of augmented pressure (AP) and pulse pressure (PP), AP/PP [28]. PWVcf is obtained by dividing the distance from the carotid pulse and femoral pulse as measured by a tape measure by the time taken for the arterial pulse to propagate to the carotid and femoral arteries [62].

Arterial stiffness, which is an indicant of the stiffening of the large arteries, increases with age, especially in central arteries, such as the aorta and carotid artery. Arterial stiffness has also been found to increase with overweightness and obesity, which leads to an increased risk of arteriosclerosis [63]. Overweight and obese children have been found to have high arterial stiffness, assessed through pulse wave velocity [64]. Similar findings were also found in young overweight individuals who had higher arterial stiffness levels compared to the normal weight individuals [62]. Hemodynamic changes accompanying overweightness and obesity include increased arterial wall stress, smooth muscle cell proliferation, thickening of vessel walls, and eventually reduced arterial compliance/increased arterial stiffness. Thickening of vessel walls is likely caused by increased total blood volume and cardiac output to compensate for the metabolic requirements of excess fat [65, 66]. Alterations of hemodynamics, together with other markers of obesity, including chronic inflammation and endothelial dysfunction, have been shown to contribute to the impairment of vasculature structure and function in obese individuals [67]. Thus, the development of arterial stiffness seems to be due to coping mechanisms by the body to meet the metabolic demands caused by excess fat gain. In overweight and obese children, this mechanism seems to develop very early on and carries on until adolescence and adulthood if interventions to reverse this condition are not introduced.

3. Health benefits of aerobic, interval sprinting, and resistance exercise

The health benefits of regular aerobic exercise are well known. Physical activity has been ranked as one of the leading health interventions, which is used to reduce sedentary behavior in children, adolescents, and adults [68]. Currently, exercise guidelines include 150 minutes of exercise per week that consist of moderate and vigorous physical activity combined with resistance training. It has been shown that performing 150 minutes of regular exercise per week results in a reduction of mortality risk by 30% and a decreased risk of diabetes, cancer, depression, and stroke [69]. Despite the positive benefits of regular physical activity, however, many people do not comply with the minimum exercise requirements to maintain health. Lack of motivation and time constrains are possibly the underlying reasons of not performing regular exercise. The importance of regular physical activity is so overwhelming that exercise now has been regarded as medicine.

Exercise has been widely used as preventative medicine to reduce the risk and incidence of cardiovascular and metabolic diseases related to sedentary and unhealthy living. Regular exercise has been shown to improve health and reduce the severity of diseases accompanying an unhealthy lifestyle. The benefits of exercise are overwhelming, and it has been shown that exercise can be used therapeutically for conditions, such as hypertension and insulin resistance [70], dyslipidemia [71, 72], type 2 diabetes, obesity [73], and endothelial dysfunction [74]. Exercise can also be used to improve cardiovascular and metabolic dysfunction that includes enhancing adipokine, cardiometabolic, and other clinical markers [75]. Thus, it is clear that exercise can be used as therapeutic or preventive medicine in order to alleviate lifestyle diseases.

The types of exercise used in past research include aerobic exercise (continuous walking, jogging, and cycling), high-intensity interval training (HIIT), and resistance training (e.g., weights). A study showed that either supervised or unsupervised aerobic exercise resulted in a reduction of body mass index in overweight and obese adolescents [76]. Whereas other studies have shown that moderate endurance training [77] and interval sprinting exercise [12] have a positive effect on HRV. Both HIIT and aerobic training have been used to induce improvement in HRV in different populations, such as young overweight individuals, type 2 diabetic patients, and older adults. Twenty minutes of HIIT on a stationary cycle ergometer, proceeded with 5 minutes of warm-up and 5 minutes of cool-down, three times per week, for 12 weeks have been found to improve parasympathetic activity [12] in young overweight males (**Figure 3**).

A study examining type 2 diabetic individuals has also found similar results when 30 minutes, four times per week for 12 weeks of HIIT training on a treadmill resulted in improvement in HRV by 19% [78]. This form of exercise consisted of 3 minutes of warm-up, 6×2 minutes of high intensity at 80–90% of heart rate maximum, separated by 6 minutes of moderate intensity at 50–60% of heart rate maximum with 2-minute recovery intervals, and 3 minutes of cool-down. The improvement of HRV in type 2 diabetes is significant as exercise could potentially prevent type 2 diabetic patients progressing to a condition called diabetic neuropathy later in life. Another HIIT study also showed that HRV was significantly improved in older individuals, whose average age was 74 years, following 14 weeks of cycle ergometer HIIT exercise [79]. Thus, both continuous aerobic exercise and HIIT had positive effects on autonomic function by increasing HRV levels. However, HIIT is currently regarded as a type of exercise that is superior to a typical continuous moderate intensity of aerobic exercise in terms of time efficiency and clinical benefits. The effectiveness of HIIT has been well established in youth [80] and in overweight adult men [12] and women [13]. Twelve weeks of HIIT have shown to improve cardiovascular function, physical fitness, assessed through a maximal oxygen uptake test, and body composition in young overweight women [81]. It seems that both continuous steady state aerobic exercise and HIIT are beneficial for health.

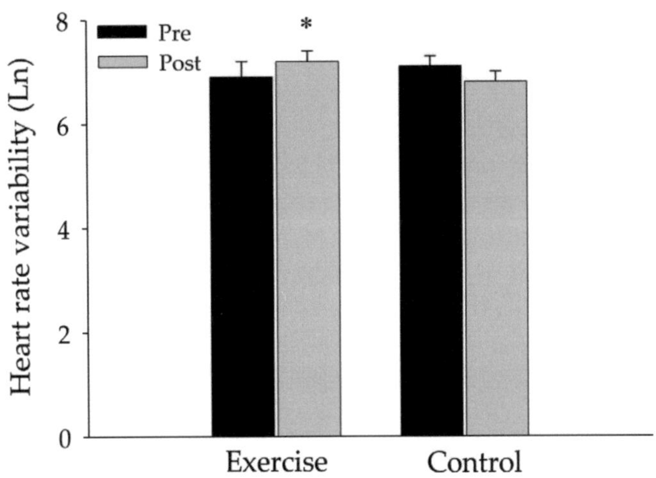

Figure 3. Heart rate variability of overweight young males at pre and post 12 weeks of interval sprinting exercise. *Significant difference between groups, *P*<0.05.

Modification of HIIT protocols, however, may be needed to suit different populations. For example, HIIT to induce athlete performance would be different with the HIIT used to induce cardiovascular health in healthy sedentary or diseased individuals. Certain exercise effects or adaptations that occur following HIIT training may not occur or be apparent after regular aerobic exercise. Depending on health markers and conditions examined, the magnitude of change following aerobic exercise may be smaller than that of HIIT. For example, it has been demonstrated that arterial stiffness, assessed through PWV [82] and autonomic function [82] are normalized following HIIT in hypertensive individuals, but not following continuous moderate exercise. With HIIT training, the exercise drop-out rate was found to be less [12, 83] compared to continuous steady state aerobic exercise. HIIT is also deemed to be superior in terms of improvement in cardiovascular health compared to regular aerobic exercise [84]. Thus, HIIT may be needed to be incorporated into daily life to induce extra health benefits. HIIT defined as repeated bouts of high-intensity exercise interspersed by rest for 20–30 minutes has also been used to prevent or to reduce severity of diseases related to unhealthy lifestyles. HIIT has been found to improve cardiac and metabolic health of young overweight males [12] and females [83]. The AIx was found to be reduced by 4% (**Figure 4**), whereas PWVcf velocity was also reduced by 0.4 m.s^{-1} (**Figure 5**) following 12 weeks of HIIT in young overweight males [12].

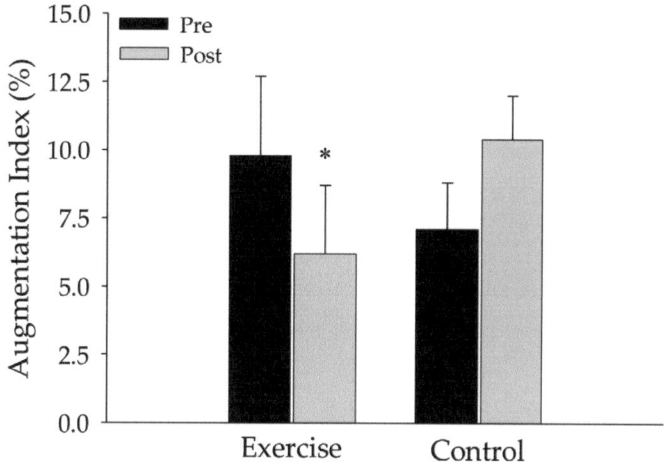

Figure 4. Augmentation index of young overweight males at pre and post 12 weeks of interval sprinting exercise. *Significant difference between groups, $P<0.05$.

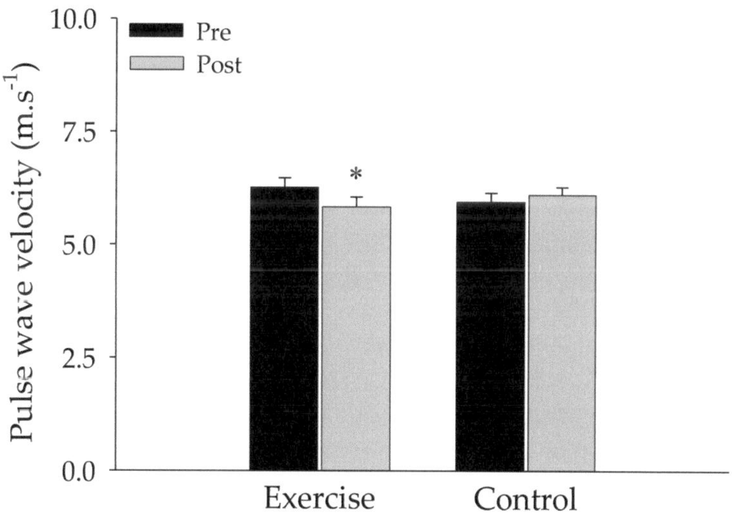

Figure 5. Pulse wave velocity of young overweight males at pre and post 12 weeks of interval sprinting exercise. *Significant difference between groups, $P<0.05$.

The HIIT employed was an 8-s pedaling sprint at a cadence of 100–120 revolutions per minute (rpm) at 0.5–1 kg of load, followed by a period of lighter intensity exercise at a cadence of 30–40 rpm for 12 seconds, repeated for 20 minutes. Another type of HIIT included cycling on a stationary bike at 80–85% of maximal oxygen uptake for 4 minutes with 5-minute rest intervals, repeated six to eight times [85]. HIIT has also been used in a number of clinical studies involving cardiac rehabilitation, chronic obstructive pulmonary disease, and intermittent claudication disease patients. Thus, a range of interval training exercise programs have been employed to improve cardiovascular and metabolic health.

Resistance exercise is another type of exercise that contributes to cardiovascular and metabolic health. Participating in resistance exercise can maintain muscle mass that declines with aging. Acute resistance training has been found to reduce systolic blood pressure by 11 mmHg and mean arterial pressure by 12 mmHg and systolic blood pressure by 13 mmHg and mean arterial pressure by 12 mmHg with 40 and 80% maximum weight, respectively [86]. The mechanisms underlying this reduction in blood pressure following resistance training is thought to be due to an increased blood flow and shear stress that act on vascular endothelial cells. The increased muscle contraction leads to an increased production of nitric oxide, an important vasodilator [87]. It appears that this mechanism occurs independently of the exercise intensity employed [88].

The effect of resistance training on arterial stiffness, however, is equivocal. Several studies [89, 90] have shown an unfavorable effect of resistance training on arterial stiffness, whereas others have shown no alteration in arterial stiffness [82, 91]. A reduction in central (increased arterial stiffness) but not peripheral arterial compliance was found following 4 months of resistance training in young and healthy middle-aged men [89]. However, only brachial artery endothelial function was improved following 1 year of resistance training, but body mass index, body composition, blood lipids, and lean muscle mass improved [92]. A meta-analysis [91] found that young adults had their arterial stiffness elevated from 14.3 to 20.1% following high-intensity resistance training. In contrast, it has been shown that progressive high-intensity resistance training without an increase in training volume did not alter arterial stiffness in young individuals [90]. Interestingly, the association between resistance training and arterial stiffness was not found in middle-aged individuals [90]. Although high-intensity resistance training has been found to increase arterial stiffness by 11.6%, moderate intensity resistance training did not seem to induce the same effect [90]. Another study showed an unexpected finding when improvement in the muscular strength of young individuals was inversely correlated with arterial stiffness [93]. This finding suggests that resistance training attenuates arterial stiffness. Different protocols and populations possibly could have contributed to the variability of results. Further studies looking at resistance training and arterial stiffness need to be carried out. Overall, regular HIIT is recommended, especially for people with time constraints. A combination of aerobic exercise and resistance training is also highly recommended as muscle wasting could occur with aging and a sedentary lifestyle.

4. Conclusion

A sedentary lifestyle leads to an increased risk of cardiovascular disease and dysfunction, such as high blood pressure, stiffening of the arteries, cardiac autonomic dysfunction, and metabolic dysfunction. If children become overweight or obese, these conditions described above appear to develop earlier. Lifestyle modification for children to avoid childhood obesity is paramount to further reduce the risk of health complications later in life. Therapeutic drug interventions can be introduced; however, overwhelming evidence suggests that regular exercise has been shown to prevent and to reduce the severity of disease-related lifestyle. Exercise is a powerful stimulus that can reduce and prevent the occurrence of cardiovascular and metabolic dysfunction. Therefore, exercise indeed is medicine that can improve the quality of life. Both regular aerobic exercise and HIIT are beneficial for health. HIIT, in particular, is highly recommended, especially for people with time constraints, and HIIT combined with resistance training is potentially highly beneficial. People of all ages should be encouraged to incorporate regular exercise into their daily lifestyle.

Author details

Yati N. Boutcher

Address all correspondence to: y.boutcher@unsw.edu.au

School of Medical Sciences, Faculty of Medicine, University of New South Wales, Australia, Sydney, Australia

References

[1] Lee I-M, Shiroma EJ, Lobelo F, Puska P, Blair SN, Katzmarzyk PT. Effect of physical inactivity on major non-communicable diseases worldwide: an analysis of burden of disease and life expectancy. Lancet. 2012;380(9838):219-229 doi:10.1016/S0140-6736(12)61031-9.

[2] Kohl HW 3rd, Craig CL, Lambert EV, Inoue S, Alkandari JR, et al. The pandemic of physical inactivity: global action for public health. Lancet. 2012;380(9838):294-305 doi: 10.1016/S0140-6736(12)60898-8.

[3] Antoniades O, Helen T. Douda HT, Papazoglou D, Tokmakidis SP. Prevalence of hypertension and determinants of cardiac function in overweight and obese children and adolescents. J Phy Activ Nutr Rehab. PANR e-ISSN: 2016;2421-7824.

[4] Davison K, Bircher S, Hill A, Coates AM, Howe PR, Buckley JD. Relationships between obesity, cardiorespiratory fitness, and cardiovascular function. J Obes. 2010;2010:191253. doi:10.1155/2010/191253.

[5] Poirier P, Giles TD, Bray GA, Hong Y, Stern JS, et al. Obesity and cardiovascular disease: pathophysiology, evaluation, and effect of weight loss: an update of the 1997 American Heart Association Scientific Statement on Obesity and Heart Disease from the Obesity Committee of the Council on Nutrition, Physical Activity, and Metabolism. American

Heart Association; Obesity Committee of the Council on Nutrition, Physical Activity, and Metabolism. Circulation. 2006;113(6):898-918.

[6] Poirier P, Giles TD, Bray GA, Hong Y, Stern JS, et al. Obesity and cardiovascular disease: pathophysiology, evaluation, and effect of weight loss. Arterioscler Thromb Vasc Biol. 2006;26(5):968-976.

[7] Poirier P, Martin J, Marceau P, Biron S, Marceau S. Impact of bariatric surgery on cardiac structure, function and clinical manifestations in morbid obesity. Expert Rev Cardiovasc Ther. 2004;2:193-201.

[8] de Divitiis O, Fazio S, Petitto M, Maddalena G, Contaldo F, Mancini M. Obesity and cardiac function. Circulation. 1981;64(3):477-482.

[9] Boutcher YN, Hopp JP, Boutcher SH. Acute effect of a single bout of aerobic exercise on vascular and baroreflex function of young males with a family history of hypertension. J Hum Hypertens. 2011;25(5):311-319. doi:10.1038/jhh.2010.62.

[10] Goldberg MJ, Boutcher SH, Boutcher YN. The effect of 4 weeks of aerobic exercise on vascular and baroreflex function of young men with a family history of hypertension. J Human Hypertens. 2012;26;11:644-649.

[11] Blair SN. Physical inactivity: the biggest public health problem of the 21st century.Br J Sports Med. 2009;43(1):1-2

[12] Heydari M, Boutcher YN, Boutcher SH. High-intensity intermittent exercise and cardiovascular and autonomic function. Clin Auton Res. 2013;23:57-65.

[13] Dunn SL, Siu W, Freund J, Boutcher SH. The effect of a lifestyle intervention on metabolic health in young women. Diabetes Metab Syndr Obes. 2014;19(7):437-444.

[14] Heidenreich PA, Trogdon JG, Khavjou OA, Butler J, Dracup K, et al. Forecasting the future of cardiovascular disease in the United States: a policy statement from the American Heart Association. Circulation. 2011;123:933-944.

[15] Tanaka H, Dinenno FA, Monahan KD, Clevenger CM, DeSouza CA, et al. Aging, habitual exercise, and dynamic arterial compliance. Circulation. 2000;102:1270-1275.

[16] Monahan KD, Dinenno FE, Seals DR, Clevenger CM, Desouza CA, et al. Age-associated changes in cardiovagal baroreflex sensitivity are related to central arterial compliance. Am J Physiol Heart Circ Physiol. 2001; 281: H284-H289.

[17] Del Colle S, Milan A, Caserta M, Dematteis A, Naso D, et al. Baroreflex sensitivity is impaired in essential hypertensives with central obesity. J Hum Hypertens. 2007;21(6): 473-478.

[18] Grassi G, Seravalle G, Colombo M, Bolla G, Cattaneo BM, et al. Body weight reduction, sympathetic nerve traffic, and arterial baroreflex in obese normotensive humans. Circulation. 1998;97:2037-2042.

[19] Pedersen BK, Saltin B. Evidence for prescribing exercise as therapy in chronic disease. Scand J Med Sci Sports. 2006;16(Suppl. 1):3-63. doi:10.1111/j.1600-0838.2006.00520.x

[20] Lee I-M, Shiroma EJ, Lobelo F, Puska P, Blair SN, Katzmarzyk PT. Effect of physical inactivity on major non-communicable diseases worldwide: an analysis of burden of disease and life expectancy. Lancet. 2012;380(9838):219–229 doi:10.1016/S0140-6736(12)61031-9.

[21] Metzl JD, Heffernan A. The exercise cure: a doctor's all-natural, no pill prescription for better health & longer life. Emmaus, PA: Rodale; 2013.

[22] Cardinal BJ, Park EA, Kim M, Cardinal MK. If Exercise is Medicine, Where is Exercise in Medicine? Review of US Medical Education Curricula for physical activity-related content. J Phys Act Health. 2015;12(9):1336-1343. doi:10.1123/jpah.2014-0316.

[23] WHO, Global Health Observatory data 2010-2014. Prevalence of overweight and obesity age 18+ 2010-2014, age standardized estimate (Internet) 2016. Available from: http://www.who.int/gho/publications/world_health_statistics/en/ [Accessed: 2016-06-15].

[24] World Health Organization (1999) Obesity: preventing and managing the global epidemic: report of a WHO consultation.

[25] WHO technical report series: 894. Geneva. World Health Organization (2002) The European Health report 2002. WHO regional publications. European series; No. 97.Kopenhagen.

[26] Fang J, Wylie-Rosett J, Cohen HW, Kaplan RC, Alderman MH. Exercise, body mass index, caloric intake, and cardiovascular mortality. Am J Prev Med. 2003; 25(4):283-289. doi:10.1016/S0749-3797(03)00207-1.

[27] Wei M, Kampert JB, Barlow CE, Nichaman MZ, Gibbons LW, Paffenbarger RS Jr, et al. Relationship between low cardiorespiratory fitness and mortality in normal-weight, overweight, and obese men. JAMA. 1999;282(16):1547-1553. doi:10.1001/jama.282.16.1547.

[28] Lakka HM, Laaksonen DE, Lakka TA, et al. The metabolic syndrome and total and cardiovascular disease mortality in middle-aged men. WHO Guidelines. JAMA. 2002;228:2709-2716.

[29] Quinones MJ, Hernandez-Pamplona M, Schelbert H, Bulnes-Enriquez I, Jimenez X et al. Coronary vasomotor abnormalities in insulin-resistant individuals. Ann Intern Med. 2004; 140: 700-708.

[30] Makimattila S, Yki-Jarvinen H. Endothelial dysfunction in human diabetes. Curr Diab Rep. 2002;2:26-36.

[31] Weiss R, Dziura J, Burgert TS, Tamborlane WV, Taksali SE, et al. Obesity and the metabolic syndrome in children and adolescents. N Engl J Med. 2004; 350:2362-2374.

[32] Kaufman CL, Kaiser DR, Steinberger J, Kelly AS, Dengel DR. Relationships of cardiac autonomic function with metabolic abnormalities in childhood obesity. Obesity (Silver Spring). 2007;15(5):1164-1171.

[33] Dietz WH. Overweight in childhood and adolescence. N Engl J Med. 2004;350:855-857.

[34] Cambuli VM, Musiu MC, Incani M, Paderi M, Serpe R, et al. Assessment of adiponectin and leptin as biomarkers of positive metabolic outcomes after lifestyle intervention in overweight and obese children. J Clin Endocrinol Metab. 2008; 93(8):3051-3057. doi: 10.1210/jc.2008-0476.

[35] Evrengul H, Tanriverdi H, Kose S, Amasyali B, Kilic A. The relationship between heart rate recovery and heart rate variability in coronary artery disease. Ann Noninvasive Electrocardiol. 2006;11(2):154-162.

[36] Quilliot D, Zannad F, Ziegler O. Impaired response of cardiac autonomic nervous system to glucose load in severe obesity. Metabolism. 2005;54(7):966-974.

[37] Windham BG, Fumagalli S, Ble A, Sollers JJ, Thayer JF, et al. The relationship between heart rate variability and adiposity differs for central and overall adiposity. J Obes. 2012; 1–8. doi:10.1155/2012/149516.

[38] Christou DD, Jones PP, Pimentel AE, Seals DR (2004) Increased abdominal-to-peripheral fat distribution contributes to altered autonomic-circulatory control with human aging. Am J Physiol. 2004; 287(4): H1530-H1537.

[39] Task Force European Society Cardiology of the North American Society Pacing and Electrophysiology. Heart rate variability: standards of measurement, physiological interpretation, and clinical use. Circulation. 1996;93(5):1043-1065.

[40] Bigger JT Jr, Fleiss JL, Rolnitzky LM, Steinman RC Frequency domain measures of heart period variability to assess risk late after myocardial infarction. J Am Coll Cardiol. 1993;21(3):729-736.

[41] Kleiger RE, Miller JP, Bigger JT Jr, Moss AJ. Decreased heart rate variability and its association with increased mortality after acute myocardial infarction. Am J Cardiol. 1987;59(4):256-262. doi:10.1016/0002-9149(87)90795-8.

[42] Fraley MA, Birchem JA, Senkottaiyan N, Alpert MA. Obesity and the electrocardiogram. Obes Rev. 2005;6(4):275-281. doi:10.1111/j.1467-789X.2005.00199.x.

[43] Ramaekers D, Ector H, Aubert AE, Rubens A, Van de Werf F Heart rate variability and heart rate in healthy volunteers. Is the female autonomic nervous system cardioprotective? Eur Heart J. 1998;19(9):1334-1341. doi:10.1053/euhj.1998.1084.

[44] La Rovere MT, Mortara A, Schwartz PJ. Baroreflex sensitivity. J Cardiovasc Electrophysiol. 1995;6:761-774.

[45] Martini G, Riva P, Rabbia F, Molini V, Ferrero GB, et al. Heart rate variability in childhood obesity. Clin Auton Res. 2001;11:87-91.

[46] Nagai N, Matsumoto T, Kita H, Moritani T. Autonomic nervous system activity and the state and development of obesity in Japanese school children. Obes Res. 2003;11:25-32.

[47] Rabbia F, Silke B, Conterno A, Grosso T, De Vito B, et al. Assessment of cardiac autonomic modulation during adolescent obesity. Obes Res. 2003;11:541-548.

[48] Riva G, Martini G, Milan A, Paglieri C, Chiandussi L, et al. Obesity and autonomic function in adolescence. Clin Exp Hypertens. 2001;23:57-67.

[49] Yakinci C, Mungen B, Karabiber H, Tayfun M, Evereklioglu C. Autonomic nervous system functions in obese children. Brain Dev. 2000;22:151-153.

[50] Beauloye V, Zech F, Tran HT, Clapuyt P, Maes M, et al. Determinants of early atherosclerosis in obese children and adolescents. J Clin Endocrinol Metab. 2007;92:3025-3032

[51] Berenson GS, Srinivasan RS, Bao W, Newman WP 3rd, Tracy RE, et al. Association between multiple cardiovascular risk factors and atherosclerosis in children and young adults. N Engl J Med. 1998;338:1650-1656.

[52] Reinehr T, Kratzsch J, Kiess W, Andler W. Circulating soluble leptin receptor, leptin, and insulin resistance before and after weight loss in obese children. Int J Obes Relat Metab Disord. 2005;29:1230-1235.

[53] Steinberger J, Steffen L, Jacobs DR, Moran A, Hong C-P, et al. Relation of leptin to insulin resistance syndrome in children. Obes Res. 2003;11:1124-1130.

[54] Lihn AS, Pedersen SB, Richelsen B. Adiponectin: action, regulation and association to insulin sensitivity. Obes Rev. 2005;6(1):13-21.

[55] Ross R. Atherosclerosis—an inflammatory disease. N Engl J Med. 1999;340:115-126

[56] Mano T, Masuyama T, Yamamoto K, Naito J, Kondo H et al. Endothelial dysfunction in the early stage of atherosclerosis precedes appearance of intimal lesions assessable with intravascular ultrasound. Am Heart J. 1996;131:231-238.

[57] Tooke JE. Microvascular function in human diabetes: a physiologic perspective. Diabetes. 1995;44:721-726.

[58] Gates PE, Seals DR. Decline in large elastic artery compliance with age: a therapeutic target for habitual exercise. Br J Sports Med. 2006;40(11): 897-899.

[59] Dengo AL, Dennis EA, Orr JS, Marinik EL, Ehrlich E et al. Arterial destiffening with weight loss in overweight and obese middle-aged and older adults. Hypertension. 2010; 55:855–861.

[60] Vlachopoulos C, Aznaouridis K, Stefanadis C. Prediction of cardiovascular events and all-cause mortality with arterial stiffness: a systematic review and meta-analysis. J Am Coll Cardiol. 2010; 55:1318-1327.

[61] Ikonomidis I, Mazarakis A, Papadopoulos C, Patsouras N, Kalfarentzos F et al. Weight loss after bariatric surgery improves aortic elastic properties and left ventricular

function in individuals with morbid obesity: a 3-year follow-up study. J Hypertens. 2007; 25:439-447.

[62] Chandrakumar D, Boutcher SH, Boutcher YN. Acute exercise effects on vascular and autonomic function in overweight men. J Sports Med Phys Fitness. 2015;55;1-2:91-102.

[63] Huang HM, Li S, Xu H. [Association between overweight, obesity and arterial stiffness in community residents]. Zhonghua Xin Xue Guan Bing Za Zhi. 2011; 39(10):950-954.

[64] Pandit D, Kinare A, Chiplonkar S, Khadilkar A, Khadilkar V. Carotid arterial stiffness in overweight and obese Indian children. J Pediatr Endocrinol Metab. 2011;24(1-2):97-102.

[65] Zebekakis PE, Nawrot T, Thijs L, Balkestein EJ, van der Heijden-Spek J, et al. Obesity is associated with increased arterial stiffness from adolescence until old age. J. Hypertens. 2005;23:1839-1846.

[66] De Michele M, Panico S, Iannuzzi A, Celentano E, Ciardullo AV, et al. Association of obesity and central fat distribution with carotid artery wall thickening in middle-aged women. Stroke. 2002;33:2923-2928.

[67] Stapleton PA, James ME, Goodwill AG, Frisbee JC. Obesity and vascular dysfunction. Pathophysiology. 2008;15:79-89.

[68] Heath GW, Kolade VO, Haynes JW. Exercise is Medicine™: A pilot study linking primary care with community physical activity support. Prev Med Rep. 2015; 2:492-497.

[69] MacIntosh BF, Yungblut S, Frankovich F, Oh P, Fowles J. Exercise is Medicine: Canada: engaging patients in physical activity dialogue-practical tools, CME workshop to assist physicians. Ontario Medical Review, March 2016;30-33.

[70] Dengel DR, Galecki AT, Hagberg JM, Pratley RE. The independent and combined effects of weight loss and aerobic exercise on blood pressure and oral glucose tolerance in older men. Am J Hypertens. 1998;11:1405-1412.

[71] Gondim OS, de Camargo VT, Gutierrez FA, Martins PF, Passos ME, et al. Benefits of regular exercise on inflammatory and cardiovascular risk markers in normal weight, overweight and obese adults. PLoS One. 2015;10(10):e0140596. doi:10.1371/journal.pone.0140596. eCollection 2015.

[72] Linhart A, Gariepy J, Giral P, Levenson J, Simon A. Carotid artery and left ventricular structural relationship in asymptomatic men at risk for cardiovascular disease. Atherosclerosis. 1996;127:103-112.

[73] Zinman B, Vranic M. Diabetes and exercise. Med Clin North Am. 1985;69:145–157.

[74] Niebauer JJ, Cooke P. Cardiovascular effects of exercise: role of endothelial shear stress. J Am Coll Cardiol. 1996;28:1652-1660.

[75] Mendham AE, Duffield R, Marino F, Coutts AJ. A 12-week sports-based exercise programme for inactive Indigenous Australian men improved clinical risk factors associated with type 2 diabetes mellitus. J Sci Med Sport. 2015;18(4):438-443. doi: 10.1016/j.jsams.2014.06.013

[76] Ruotsalainen H, Kyngäs H, Tammelin T, Kääriäinen M. Systematic review of physical activity and exercise interventions on body mass indices, subsequent physical activity and psychological symptoms in overweight and obese adolescents. J Adv Nurs. 2015;71(11):2461-2477. doi:10.1111/jan.12696.

[77] Hottenrott K, Hoos O, Esperer HD. Heart rate variability and physical exercise. Herz. 2006;31(6):544–552. doi:10.1007/s00059-006-2855-1

[78] Parpa KM, Michaelides MA, Brown BS. Effect of high intensity interval training on heart rate variability in individuals with type 2 diabetes J Exerc Physiol. 2009; 12;4: 23-29.

[79] Pichot V, Roche F, Denis C, Garet M, Duverney D, et al. Interval training in elderly men increases both heart rate variability and baroreflex activity. Clin Auton Res. 2005;15(2): 107-115.

[80] García-Hermoso A, Cerrillo-Urbina AJ, Herrera-Valenzuela T, Cristi-Montero C, Saavedra JM, et al. Is high-intensity interval training more effective on improving cardiometabolic risk and aerobic capacity than other forms of exercise in overweight and obese youth? A meta-analysis. Obes Rev. 2016;17(6):531-540. doi:10.1111/obr.12395.

[81] Sijie T, Hainai Y, Fengying Y, Jianxiong W. High intensity interval exercise training in overweight young women. J Sports Med Phys Fitness. 2012;52(3):255-262.

[82] Rakobowchuk M, Harris E, Taylor A, Cubbon RM, Birch KM (2013) Moderate and heavy metabolic stress interval training improve arterial stiffness and heart rate dynamics in humans. Eur J Appl Physiol. 2013;113:839-849.

[83] Dunn SL, Siu W, Freund J, Boutcher SH. The effect of a lifestyle intervention on metabolic health in young women. Diabetes Metab Syndr Obes. 2014;19(7):437-444.

[84] Ciolac EG. High-intensity interval training and hypertension: maximizing the benefits of exercise? Am J Cardiovasc Dis. 2012;2(2):102-110.

[85] Boutcher YN, Hamer M, Boutcher SH. The effect of acute plasma volume expansion on venous capacitance. J Sports Med Phys Fitness. 2003;43(1):105-110.

[86] Cavalcante PA, Rica RL, Evangelista AL, Serra AJ, Figueira A Jr, et al. Effects of exercise intensity on post exercise hypotension after resistance training session in overweight hypertensive patients. Clin Interv Aging. 2015;18(10):1487-1495. doi:10.2147/CIA.S79625. eCollection 2015.

[87] Goto C, Higashi Y, Kimura M, Noma K, Hara K, et al. Effect of different intensities of exercise on endothelium-dependent vasodilation in humans: role of endothelium-dependent nitric oxide and oxidative stress. Circulation. 2003;108(5):530-535.

[88] Bentes CM, Costa PB, Neto GR, Costa e Silva GV, de Salles BF, et al. Hypotensive effects and performance responses between different resistance training intensities and exercise orders in apparently health women. Clin Physiol Funct Imaging. 2014; doi: 10.1111/cpf.12144.

[89] Miyachi M, Kawano H, Sugawara J, Takahashi K, Hayashi K, et al. Unfavorable effects of resistance training on central arterial compliance: a randomized intervention study. Circulation. 2004;110(18):2858-2863.

[90] Casey DP, Beck DT, Braith RW. Progressive resistance training without volume increases does not alter arterial stiffness and aortic wave reflection. Exp Biol Med. 2007; 232;9:1228-1235.

[91] Miyachi M. Effects of resistance training on arterial stiffness: a meta-analysis. Br J Sports Med. 2013;47(6):393-396.

[92] Olson TP, Dengel DR, Leon AS, Schmitz KH. Moderate resistance training and vascular health in overweight women. Med Sci Sports Exerc. 2006;38(9):1558-1564.

[93] Fahs CA, Heffernan KS, Ranadive S, Jae SY, Fernhall B. Muscular strength is inversely associated with aortic stiffness in young men. Med Sci Sports Exerc. 2010; 42(9):1619-1624.

Chapter 2

Body Shape, Image, and Composition as Predictors of Athlete's Performance

Arnulfo Ramos-Jiménez, Rebeca Chávez-Herrera,
Aida S. Castro-Sosa, Lilia C. Pérez-Hernández,
Rosa P. Hernández Torres and David Olivas-Dávila

Additional information is available at the end of the chapter

http://dx.doi.org/10.5772/65034

Abstract

Body shape, image, and composition are three different but related concepts used to describe people. Body shape, also known as somatotype, represents the tangible body, which can be externally observed and measured without destroying or hurting it. On the contrary, body image represents the subjective and intangible human nature, a construct that we try to define by applying validated scientific tools—a set of dimensions easily affected by psychological perception. Instead, body composition represents the physical parts grouped into similar compartments. Due to the fact that it cannot be observed or measured with the naked eye, and in order to reduce measurement error, we try to measure them with the highest and most accurate available technology. Shape, image, and composition affect sport's performance. Sports literature mentions, sometimes interchangeably, form, image, and body composition. So when we refer to them we have to distinguish them. Social, political, economic, cultural, educational, and genetic factors influence them. Technological advances in determining the shape and composition are reliable, but not the ones for body image, which needs further development. In this paper, the interrelation of these three aspects is described, with health and sport's ambit indicators.

Keywords: anthropometry, somatotype, physical exercise

1. Introduction

Humankind has always tried to understand what surrounds it through characterization and division of objects into different groups, that is to say we classify them according to their particular characteristics of shape, size, color, texture, smell, weight, hardness, density, resistance, flexibility, temperature, etc. Sociopolitical and cultural sciences try to understand mankind though its history, culture, education, socioeconomic status, politics, religion, migration, etc. In the biomedical sciences, we strive to integrate many of these classifications in order to know the individual and social behavior of the human race, their preferences, feelings, and sufferings. Mankind with all of this creates a great quantity of sciences to theoretically facilitate, make more comfortable, and longer its existence.

With the purpose of understanding body structures, how they are organized and affect people, physical and functional anthropology, macroscopic anatomy, and basic and applied psychology intertwine, creating sometimes other sciences, including kinanthropometry. It is within this set of sciences where the present work is carried out.

From the point of view of macroscopic and functional anatomy, the human body consists of different types of cells, grouped and compartmentalized to form tissues, organs, and systems. Said division has allowed kinanthropometry and other sciences to regroup them in different macroscopic structures called components, including bone mass, lean mass, visceral mass, muscle mass, and fat mass. The amount, proportion, and distribution of these components give the body a particular physical form [1]; intervening around these a great number of disciplines and measuring instruments in order to determine them. In nonpathological conditions, the variability of these components between people depends on several factors, among the main ones: genetic aspects, gender, hormonal, pregnancy-related, age, physical activity, physical postures, environmental, and cultural. All of these factors and natural selection have allowed the human body to adapt to its environment [2–6]. In this sense, the literature talks about body shape, body composition, and body image, which are a construct that even though it seems easy to define its description is complex. Body shape would be the objective and technically measurable representation of the exterior of the human body, which are the weight, length, diameter, circumference, and volumes of its various segments. Body composition would be the magnitude and proportion of the different body components mentioned above that in an indirect manner and *in vivo* we try to know. Body image would be the mental and subjective representation of body shape and body composition, in terms of size, shape, proportions, color, texture, position, etc. [7]. This subjective representation of the body, which depends on personal experiences, stems from the consistent and objective view we have of it; however, it is transformed or modified by various factors that will be mentioned. Body shape, composition, and image have great health and social implications. Health implications are those that affect in a positive or negative manner the health of the individual, whereas social implications are those that allow us to assertively or erroneously communicate and interact with others.

The literature reports several methods destined to determine the amount, proportion and distribution of the components that provide body shape. Given the impossibility of measuring them directly on people or due to the danger of high-ionizing radiation exposure in some procedures, the validity of some methods cannot be corroborated *in vivo*. The current methods that are most often used to determine both shape and body components are: anthropometry, digital photogrammetry in 3D on level body surface, bioelectrical impedance (IB), ultrasound, densitometry, deuterated water (labeled deuterium isotope), deuterated creatinine, spectroscopy, reading magnetic resonance imaging (MRI), computed tomography (CT), and dual-energy X-ray absorptiometry (DEXA). The importance of knowing body shape and its components lies in the association of these with physical development, aging, self-esteem, physical performance, and various metabolic disorders, such as anorexia, obesity, diabetes, osteoporosis, dyslipidemias, and metabolic syndrome, among others [3, 6, 8–11].

Regarding body image, its study is mainly done through graphical representations of the whole body, segments, or specific parts, likewise through the use of self-image. Typically, these images are selected, or modified by different methods (manual or electronic) in order to achieve specific patterns, perceive distortions of the image itself, or achieve the ideal form and image of how we want to perceive ourselves or want to look like [12]; although, we also find in the literature questionnaires which offer the reader different answer options to define their body image [13]. The validity of many of these methods is under discussion, especially due to errors in the selection and administration of the instruments and the population studied; but also by errors in the analysis and interpretation of them [14].

This paper analyzes the three concepts outlined above. Which although interrelated, they are different, these are: body shape, body image, and body composition. We analyzed equally how each one relates to physical and athletic performance.

2. Theoretical framework

2.1. Body shape

Body shape is also known as somatotype. This was initially described by Matiegka in 1921 [15], who divided body structure into four basic components: bone mass, subcutaneous fat mass, muscle mass, and residual mass. Years later, body shape was retaken by the American psychologist William Herbert Sheldon [16], who from around 4000 human photographs reconstructed the somatotype, thus creating three qualitative dimensions related to the three embryonic germ layers: (1) endomorph related to the endoderm, which largely forms the digestive tract, (2) mesomorph related to the mesoderm, which gives rise to muscle mass and bone mass, and (3) ectomorph related to the ectoderm, which forms the nervous system. Theoretically according to Sheldon, endomorphs have a slow metabolism predisposed to accumulate body fat thus making them more likely to be quiet, mesomorphs have a normal metabolism but are predisposed to develop large muscles which predisposes them to be active and rough, and ectomorphs have an accelerated metabolism that makes them lose fat mass and muscle mass, presenting thin and fearful appearances [16]. Sheldon as a psychologist tried to analyze behavior through body shape, probably taking up the ideas of Joung about the types of thinking, feeling, and perception. Subsequently, these qualitative and unscientific ideas from Sheldon were taken up and perfected by Barbara Heath (Sheldon's assistant) and Lindsay Carter, who perfected Sheldon's numerical system by developing an anthropometric system to give a scientific and quantitative character to Sheldon's somatotype [17]. Currently, this anthropometric system has been retaken by the International Association for Standardization

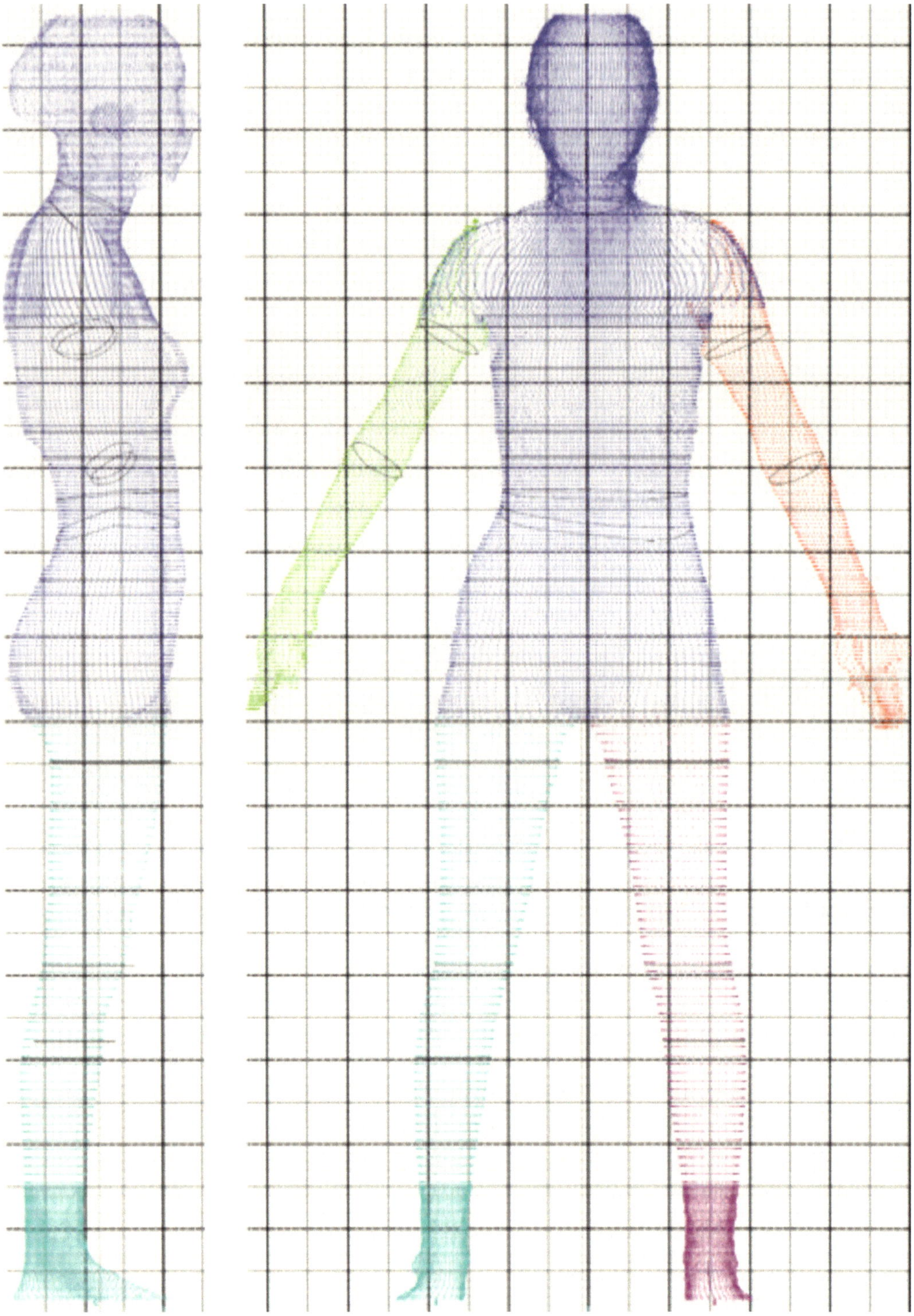

Figure 1. Body shape by Scanner in 3D (TC2-18 3D Body Scanner, USA). A small square corresponds to 1 cm^2.

of Kinanthropometry (ISAK), extending it internationally so through body measurements it will be able to determine problems with growth, development, aging, health status, as well as ergometers, and athletic-sports performance, among others (**Figure 1**).

Other authors have tried to classify body shape based on the distribution of certain body components. Bouchard [18] in the late twentieth century classified body shape based on four patterns of obesity: Pattern (1) with excess body fat and distributed in a balanced way throughout the body, pattern (2) with excess body fat and distributed mainly in the abdominal region (android type), pattern (3) with excess body fat and distributed mainly deep in the abdomen, and pattern (4) with excess body fat and distributed primarily in the gluteal muscles and thigh bone (gynaeoid type).

Differences in body shape between species, including humans, respond to the need of adjustment and survival to their environment; mainly against the weather [19], the geography [20], and food availability [21], also being present differences between genders (sexual dimorphism). Referring to the weather, and in order to maintain thermal homeostasis, individuals from cold environments are wider in comparison to thinner individuals who live in warm environments [22]. Faced with geography, wideness instead of height is also affected; however, an established pattern that indicates geography characteristics favoring the increase or decrease of corpulence has not been detected [22]. Finally, due to the technological domain of agriculture, food preservation and its ease of transportation; food availability should no longer influence morphological changes in humans [22]. However, according to the World Food Program [23], one in four children from developing countries presents stunted growth due to malnutrition. Likewise, the Food and Agriculture Organization of the United Nations [24] reports the existence of developing countries where more than 35% of the population still suffers from chronic malnutrition, including Namibia, Zambia, Central African Republic, and the Democratic People's Republic of Korea. Given the fact that malnutrition depends on the country's capacity to feed its population, morphological changes are also influenced by political and economic situations. On the other hand, due to the large number of immigrating populations, the differences due to geography and climate have begun to get lost; however, genetic, cultural, and socioeconomic patterns are those that currently dominate in the determination of body shape [25].

Currently, in a globalized world where commercial pressures are superimposed over rational judgment, the idea that being thin is favored [26], turning into a pathology the trend of extreme thinness, which leads to malnutrition, anorexia, and bulimia, especially in women [27]. Conversely, in some Arab countries (Saudi Arabia, Kuwait, and Bahrain) opposing cultural ideas on thinness exist, in which being robust is synonymous with beauty and sexiness, obesity levels being 39% for men and 50% for women [28]. However, due to interculturalization and Westernization, these young people also have the highest rates of eating disorders (~45% of young people in the case of Kuwait), who wish to be thinner [29].

2.1.1. Body shape and physical sport's performance

Body shape and physical-athletic performance are closely related, that is to say that changes in one necessarily affect the other. In addition, body shape is different both among sports and between categories and positions within a sport; each sport and its position requires from its athletes specific morphological characteristics; some dominate over others, even within each sport [30–32]. For example, football athletes are taller and leaner (meso-endomorphs), basketball players are tall and lean (meso-ectomorphic), body-builders are muscular (meso-endomorphs), long distance runners are thin, etc. [33–35], with roundness or endomorphic somatotype (synonymous with abundant fat mass), being the less fortunate feature for physical sport's performance. In this sense, being tall and lean offers competitive advantages in football [34], basketball [36], volleyball [37], soccer [38], swimming [39], and athletics [40], among others. In the same manner, Gabbett et al. [37] report that elite volleyball players tend to be taller and leaner than novice players. Rebelo et al. [38] report that soccer elite players are taller than the nonelite players. On the other hand, aerobic endurance sports, for which athletes have to mobilize their body over long distances, show that it is convenient to have small and thin bodies. For example, Moro et al. [40] reported that compared with cyclists, triathletes have lower body weight (75 vs. 79 kg) and less fat (6% vs. 8%). Arazi et al. [41] found in cross-country runners average heights of 1.75 cm and body fat percentages of 8%, with meso-ectomorphic somatotypes (1.4–4.1–3.6). Moreover, Vernillo et al. [42] found that world class Kenyan athletes display extreme sport's thinness; that is, very low endomorphy and mesomorphy and relatively high ectomorphy (balanced ectomorphy, 1.5–1.8–3.9), with body fat percentages of 4.6% (assessed by anthropometry). Comparatively, Belli et al. [43] found that mountain ultramarathon runners have meso-endomorphic (3.4–5.2–1.7) somatotypes, with fat percentages of 13%, noting that the higher endomorph content helps them obtain a good performance during their sport's execution, and withstand the physical requirements to run 217 kilometers.

Along with the previous information, systematic training over time, coupled with professionalized sports, is constructing body shape toward a mesomorphic constitution, with a slight endomorphic decrease [44].

As it has been observed, one way of analyzing both body shape and sports performance is through the somatotype. In sports that require the use of body weight as a form of attack, endomorphy exceeds five units. Mesomorphy is preferred in all other sports with values greater than 5 units, and rarely less than 3 units, especially in elite athletes [32], except for some distance runners as Kenyans. This tells us that when referring to somatotype, mesomorphy is the best predictor of physical and athletic performance [30].

2.2. Body image

Body image is a complex construct that encompasses several dimensions, including perceptions of the operation (appropriate or not) of our body, and attitudes we take to respond effectively to daily life activities; all of this according to our experience, self-awareness and self-perception [8]. In other words, it is not only the mental representation of our external physical appearance, but also the concepts, attitudes, and feelings we have about our body (height, size, proportions, weight, color, voice, posture, etc.), which allow us to interact with the environment and raise affective emotions, either of tranquility and happiness or anxiety and stress [45].

Body image develops in the early years of life [46], where, probably, parents, siblings, and relatives are the first ones to shape children's body image [47]. Given the easy access to media since the first years of life (television, video games, and the Internet), it is possible for these to also be taking part in the creation of body image at an early age [48]. Harrison et al. [49] reported when studying attitudes about obesity in children 4–6 years old, that children have preferences for being thin and have psychosocial and sports disadvantages due to being obese.

Body image is positively or negatively affected by various psychosocial factors, such as health, sport's competitions, physical exercise, and nutrition, which modify how we perceive and evaluate ourselves [50]. Its distortion among the population is variable, and differs among sex, age, ethnicity, and type of population [51–53]. Regarding gender differences among athletes, women, compared to men, have higher levels of dissatisfaction with their sport's body shape and greater eating disorders [54]; in addition, while men are more concerned about sport's performance and improving their physical abilities, women care about being thin and having less body fat, making them prone to caloric restrictions, along with the intake of dietary supplements for weight loss [55]. Studying the general population, women have also been found to have greater body image distortion than men (~10% higher, 11–72% vs. 8–61% for women and men, respectively), with some body parts being more affected than others. For example, in women, abdomen, hips, and legs are the most affected (50–71%), whereas in men it is the chest and muscle tone (32–45%). Regarding age, dissatisfaction with body image increases along with years, especially among women [51]. Olvera et al. [56] while studying Hispanic and Caucasian tweens reported that teenage boys prefer to have taller bodies while girls thinner ones. On the other hand, Ferraro et al. [57] mention that even in older adults differences between gender are observed, for example, when women were asked how they perceived their body image they answered "a little big," whereas men of the same age responded "just the right size." In the case of elderly women, this can have unhealthy results like unnecessary diets, excessive weight control, and low self-esteem [50]. Being subjected to caloric restriction at older ages can lead to malnutrition, favoring the loss of muscle mass and muscle dysfunction (both problems already present in older adults), thereby increasing the presence of infections, accidents, fractures, and a set of diseases associated to malnutrition in adulthood [58]. As for the differences between ethnic groups, separate studies in U.S. population have observed that African American women have lower self-image problems than other ethnic groups, being the order as follows: African American < Asian < Caucasian < Latin; in addition, African Americans and Caucasians see themselves as being thinner than the other groups [52, 59]. Nevertheless, studies with African American college women found that they assigned higher priority to their hair and skin tone compared to Caucasian women [60]. Hair and skin tone are outstanding features for the African American population, which means that evaluations of body image should somehow include skin color and hair. Finally, it has also been reported that women in rural areas have a more positive body self-image than those from urban areas [53].

In this paper, we consider that a healthy body image is one that being healthy is well accepted by oneself (feeling comfortable with your body), on the other hand, an unhealthy one is a negative self-image, which is not accepted even though is healthy. Treatments to improve body image are diverse, including diets, exercise, surgery, cosmetic, and psychological or psychiatric treatments

[7]. Normally, the effects on body image are favored by economic (labor), social (competition), family, colleagues, and peer pressures. In addition, they are associated with other conditions, including bulimia, anorexia, and body dysmorphic disorder [45], causing these pressures extreme thinness, especially in women, and bigorexia and anabolic steroid use in men [45].

2.2.1. *Body image and physical sport's performance*

It is recognized that the mere fact of playing a sport improves the perception of body image [61]; this is possibly because the athlete's bodies resemble more of an aesthetic ideal [62]. From the point of view of the duration of exercise, Hausenblas and Downs [63] mention that athletes have a more positive body image compared to the general population. Ginis et al. [64] observe that 8 weeks of exercise, and improvement of physical abilities, increase self-esteem ($R \sim 0.50$), with self-esteem benefits being higher in aerobic exercise types compared to anaerobic. Williams et al. [65] note that only 6 weeks of circuit weightlifting, along with increased muscle strength, reduces anxiety and improves body image. In terms of frequency, Homan et al. [66] observed that people who exercise more often exhibit higher self-esteem (appreciate their body as it is); moreover, regardless of their body shape they are probably more identified with exercise. In addition, it has even been observed that even a single session of exercise increases self-esteem in people [67]. Varnes et al. [68] in a systematic review found that most studies report higher self-esteem in athletes than in nonathletes; however, differences found between athletes of different disciplines are not consistent. For example, Latorre-Roman et al. [69] found greater body dissatisfaction and physical exercise dependence among triathlon and swimming athletes, and lower in athletics and cycling athletes, all of them involved in endurance sports. Instead, Goltz et al. [70] found no differences between sports, but they did with body fat content, where athletes with higher percentage of fat presented greater dissatisfaction with their body image. Finally, it is also recognized that body image problems are easier to discover and analyze among female athletes, than in males, so probably these problems are underassessed and underdiagnosed in men, being this one of the main reasons of stress in athletes of both sexes to maintain body weight and aesthetics [71].

With technological advancement, body image processing has been refined, currently with the use of sophisticated equipment and computer programs that facilitate their interpretation and integration. In addition to it being integrated with the third concept to be discussed in this paper called "body composition." Among the equipment that facilitates this integration we have computed magnetic resonance, high-resolution tomography, X-ray densitometry, topographic photogrammetry, and multifrequency bioelectrical impedance.

2.3. Body composition

As mentioned above, for its study, we divided the human body in different compartments. This compartmentalization is not recent; Hippocrates (460 BC) somehow had already divided the human body into four fluids or basic substances called humors: blood (affectionate), phlegm (indifferent), yellow bile (bad temper), and black bile (depressive) [72]. Later, Galen (~200 AD) took these ideas and created four other psychological classifications or temperaments associated with the first: choleric (grumpy and irritable), melancholic (analytical and quiet), sanguine (optimistic and social), and phlegmatic (relaxed and calm) [73]. However, Wiltse et al. [74] mention that formal knowledge of the human body and its compartments probably started with systematic human dissections made in the third century BC by Herophilus, known as the father of anatomy. It was not until the renaissance, after breaking the religious myths about the divinity of the body that modern anatomist Vesalius made new dissections, constructing important maps of the inside of the human body [75]. With the aid of current technological advances, we can now quantify, through precise images the interior of the human body; however, these methods are associated to some health risks. Thus, the determination of body composition is simplified by fractionating the body. For this purpose, various levels of compartmentalization are proposed: molecular (water, nitrogen, calcium, potassium, sodium, and chlorine), tissues (fat mass, bone mass, muscle mass, and residual mass), and the combination of several of these, called multicompartment models. A full explanation of these models, their considerations, and equations are found in Fosbøl et al. [76].

Currently, we quantify *in vivo*, both components of the human body and the relationships between them coupled with the changes produced by various factors. A brief summary of the documents published on body composition, as well as a brief history of methods for quantification, from Hippocrates until the late twentieth century, is presented in *History of the study of human body composition: A brief review* [77].

As shown, body composition has been well investigated. The importance of the study lies in its high correlation (R^2 0.50–0.79) with diverse health parameters [78] and sport's physical performance [79]. Due to the complexity of *in vivo* study, some authors have evaluated fresh cadavers, thereby creating equations for making estimations and clinical interpretations [80–82]. Due to the difficulty of *in vivo* study, body composition's determination remains imprecise, although their correlations can reach 0.8, when several methods of determination are being simultaneously studied [83]. This indicates a 64% of convergence between methods. Therefore, we recommend using different methods, in which each one observes the body from a different angle, to provide a more complete view of it. Nevertheless, it has been observed that when different methods are used, each one of them adds a measurement error, increasing total imprecision. However, measurement error is minimized with the use of increasingly accurate and standardized instruments.

For their technological capabilities, the most accurate methods for determining body composition are: computed tomography with 92% accuracy [84], nuclear magnetic resonance with 99% confidence [81], and DEXA between 66 and 85% reliability (for femur fractures) [85]. Other methods include air displacement plethysmography (Bod Pod) and underwater weight, which through the measurement of body density estimate with equations the body composition. The remaining methods we found have been validated with the ones that have been previously mentioned. For example, anthropometry versus DEXA with 84% correlation [86], ultrasound versus computed tomography with 45–82% correlation, ultrasound versus Bod Pod with 85% correlation [87], anthropometry versus NMR with 77% correlation, and bioimpedance versus NMR with 86% correlation [88], among others.

2.3.1. *Body composition and physical sport's performance*

Body composition has always been found to be associated to physical and athletic performance; the development of each of its components is specific to each sport. Gonzalez-Neira et al. [89] when studying anthropometrically female soccer players, found that the ones with higher weight and higher body mass index (BMI) had lower aerobic capacity (R^2 ~ 055, $p <$ 0.05). Marinho et al. [90] observed while studying Jiu-Jitsu athletes, that slight but significant decreases in body fat percentage in elite athletes versus nonelite (~2.5%, $p < 0.05$). Guiraudou and collaborators [91] found while studying Rugby players, that fat mass, rather than muscle mass, is the one that is related to sports performance ($R^2 = 0.61$, $p < 0.01$), with central body fat being a predictor of better performance. In the same manner, Belli [43] reports that in ultra-marathon runners, fat content in the lower limbs and abdomen has a negative impact on performance. In some sports, minor modifications of certain components can be crucial, both before and during competitions. For example, in contact and physical-constructivism sports, weeks before competition, athletes experience a marked weight loss (5–10% of body weight) to adjust to the competition weight [92]. This is done by decreasing their caloric intake by about 1% per week, plus dehydrating 24–48 hours before competition [93]. However, if these procedures are not well controlled, they can lead to negative effects on physical performance, especially if body fluids losses are greater than or equal to 5% of the body weight [94].

Regarding muscle mass, da Silva de Souza et al. [95] have observed in elite Brazilians body-builders' average body fat percentages of 9.6%, while muscle mass percentages of 52%, with a balanced mesomorphic somatotype (1.8–8.1–0.7). Mielgo-Ayuso et al. [96] have observed that in female volleyball players, regardless of the position they play, more muscle mass correlates with greater muscle power (R^2 0.11–0.77).

In regards to the development of muscle strength and power, both biochemical-molecular and biophysical characteristics are necessary. Biochemical aspects relate to fiber type, either slow twitch (type 1, red or oxidative) or fast twitch (type 2, white or glycolytic), where type 2 fibers are the fastest [97]. As for the biophysical aspects, we refer to length, thickness, pennation angle of muscle fibers, and insertion distance of the muscle with respect to fulcrum [97, 98], in which greater length and thickness of the muscle fiber translates to increased power and speed development [99]. On the contrary, a greater pennation angle correlates with less muscular power and speed [99]. For example, Terzis and collaborators [100], while studying hammer throw athletes, noted that elite athletes versus nonelite had greater quantity of muscle type 2 fibers and a greater cross-sectional area in the said fibers (60% vs. 49% and 66% vs. 51% for type of fibers and cross-sectional area, respectively); moreover, they found that athletic performance of these athletes is positively correlated with both the parameters ($R^2 = 0.17$ and $R^2 =$ 0.86, for fiber type and muscle cross-sectional area).

3. Conclusions

Technological and scientific advances to assess shape and body composition continue to become more reliable and valid throughout the years. Therefore, now it is time to strengthen the development of techniques and procedures for assessing body image. These three aspects, together with sports performance should be analyzed together to keep the athlete in optimal conditions, not only for competition, but also for health and quality of life. However, as it has been discussed, it is often tedious, costly, and difficult to achieve comprehensive assessment,

http://dx.doi.org/10.5772/65034

due to the fact that most athletes do not have access to the technological and scientific methods for such assessments to be performed. Because of this, we are still far from achieving maximum sporting achievements.

Author details

Arnulfo Ramos-Jiménez[1]*, Rebeca Chávez-Herrera[1], Aida S. Castro-Sosa[1], Lilia C. Pérez-Hernández[1], Rosa P. Hernández Torres[1] and David Olivas-Dávila[2]

*Address all correspondence to: aramos@uacj.mx

1 Instituto de Ciencias Biomédicas, Universidad Autónoma de Ciudad Juárez, Av. Plutarco Elías Calles y Hermanos Escobar s/n, Ciudad Juárez Chihuahua, México

2 Universidad Mayor, Santiago of Chile, Chile

References

[1] Schuna JM, Peterson CM, Thomas DM, Heo M, Hong S, Choi W, Heymsfield SB. Scaling of adult regional body mass and body composition as a whole to height: Relevance to body shape and body mass index. American Journal of Human Biology. (2015); 27(3): 372–379.

[2] De Bourdeaudhuij I, Van Dyck D, Salvo D, Davey R, Reis RS, Schofield G. Sugiyama T. International study of perceived neighbourhood environmental attributes and Body Mass Index: IPEN Adult study in 12 countries. International Journal of Behavioral Nutrition and Physical Activity. (2015); 12: 62.

[3] Heydari M, Freund J, Boutcher SH. The effect of high-intensity intermittent exercise on body composition of overweight young males. Journal of Obesity. (2012); 1–8. doi: 10.1155/2012/480467.

[4] Michael YL, Gold R, Perrin N, Hillier TA. Built environment and change in body mass index in older women. Health & Place. (2013); 22: 7–10.

[5] Shloim N, Hetherington MM, Rudolf M, Feltbower RG. Relationship between body mass index and women's body image, self-esteem and eating behaviours in pregnancy: A cross-cultural study. Journal of Health Psychology. (2015); 20(4): 413–426.

[6] Winkler TW, Justice AE, Graff M, Barata L, Feitosa MF, Chu S. The influence of age and sex on genetic associations with adult body size and shape: A large-scale genome-wide interaction study. PLoS Genet. (2015); 11(10). doi:10.1371/journal.pgen.1005378.

[7] Teixeira P, Going S, Houtkooper L, Cussler E, Metcalfe L, Blew R, Sardinha L, Lohman T. Exercise motivation, eating, and body image variables as predictors of weight control. Medicine & Science in Sports & Exercise. (2006); 38(1): 179–188.

[8] Brytek-Matera A. Exploring the factors related to body image dissatisfaction in the context of obesity. Archives of Psychiatry & Psychotherapy. (2011); 13(1): 63–70.

[9] Chang Y, Guo X, Chen Y, Guo L, Li Z, Yu S, Yang H, Sun Y. A body shape index and body roundness index: two new body indices to identify diabetes mellitus among rural populations in northeast China. BMC Public Health. (2015); 19(15): 794.

[10] Krakauer NY, Krakauer JC. A new body shape index predicts mortality hazard independently of body mass index. PLoS One. (2012); 7(7): e39504.

[11] Meyer NL, Sundgot-Borgen J, Lohman TG, Ackland TR, Stewart AD, Maughan RJ, Müller W. Body composition for health and performance: A survey of body composition assessment practice carried out by the Ad Hoc Research Working Group on Body Composition, Health and Performance under the auspices of the IOC Medical Commission. British Journal of Sports Medicine. (2013); 47(16): 1044–1053.

[12] Gardner RM. Methodological issues in assessment of the perceptual component of body image disturbance. British Journal of Psychology. (1996); 87(2): 327–337.

[13] Gardner RM, Brown DL Body image assessment: A review of figural drawing scales. Personality and Individual Differences. (2010); 48(2): 107–111.

[14] Thompson JK. The (mis)measurement of body image: Ten strategies to improve assessment for applied and research purposes. Body Image. (2004); 1(1): 7–14.

[15] Matiegka J. The testing of physical efficiency. American Journal of Physical Anthropology. (1921); 4(3): 223–230.

[16] Sheldon WH. The Varieties of Human Physique. New York: Harper and Brothers; (1940).

[17] Heath BH, Carter JEL. A modified somatotype method. American Journal of Physical Anthropology. (1967); 27: 57–74.

[18] Bouchard C. Heredity and the path to overweight and obesity. Medicine and Science in Sports and Exercise. (1991); 23(3): 285–291.

[19] Mishmar D, Ruiz-Pesini E, Golik P, Macaulay V, Clark AG, Hosseini S, Sukernik RI. Natural selection shaped regional mtDNA variation in humans. Proceedings of the National Academy of Sciences. (2003); 100(1): 171–176.

[20] Fritz SA, Bininda-Emonds OR, Purvis A. Geographical variation in predictors of mammalian extinction risk: Big is bad, but only in the tropics. Ecology Letters. (2009); 12(6): 538–549.

[21] Blackburn TM, Gaston KJ, Loder N. Geographic gradients in body size: A clarification of Bergmann's rule. Diversity and Distributions. (1999); 5(4): 165–174.

[22] Ruff C. Variation in human body size and shape. Annual Review of Anthropology. (2002); 31: 211–232.

[23] Programa Mundial de Alimentos. (2015). Datos del hambre. Encontrado el 30 de junio del 2016. Available from: https://es.wfp.org/hambre/datos-del-hambre [Accessed 30th June 2016].

[24] FAO. The State of Food Insecurity in the World 2015. The FAO Hunger Map 2015. Available from: http://www.fao.org/hunger/en/ [Accessed 30th June 2016].

[25] Delavari M, Farrelly A, Renzaho A, Mellor D, Swinburn B. Experiences of migration and the determinants of obesity among recent Iranian immigrants in Victoria, Australia. Ethnicity & Health. (2013); 18(1): 66–82.

[26] Swami V. Cultural influences on body size ideals: Unpacking the impact of westernization and modernization. European Psychologist. (2013). Advance online publication. doi: 10.1027/1016-9040/a000150.

[27] Soltis CA. Dying to be a supermodel: Can requiring a healthy BMI be fashionable. Journal of Contemporary Health Law & Policy. (2009); 26: 49.

[28] Musaiger AO. Overweight and obesity in eastern Mediterranean region: Prevalence and possible causes. Journal of Obesity. (2011); 2011: 407237.

[29] Musaiger AO, Al-Mannai M, Tayyem R, Al-Lalla O, Ali EY, Kalam F, Chirane M. Risk of disordered eating attitudes among adolescents in seven Arab countries by gender and obesity: A cross-cultural study. Appetite. (2013); 60: 162–167.

[30] Gutnik B, Zuoza A, Zuozienė I, Alekrinskis A, Nash D, Scherbina S. Body physique and dominant somatotype in elite and low-profile athletes with different specializations. Medicina. (2015); 51(4): 247–252.

[31] Perroni F, Vetrano M, Camolese G, Guidetti L, Baldari C. Anthropometric and somatotype characteristics of young soccer players: Differences among categories, subcategories, and playing position. The Journal of Strength & Conditioning Research. (2015); 29(8): 2097–2104.

[32] Ponds V, Riera J, Galilea PA, Drobnic F, Banquells M, Ruiz, O. Anthropometric characteristics, body composition and somatotype sports. Reference data of San Cugat CAR. Características antropométricas, composición corporal y somatotipo por deportes. Datos de referencia del CAR de San Cugat, 1989–2013. Apunts. Medicina de l'Esport. (2015); 50 (186): 65–72.

[33] Kim H, Lee S, Choue R. Metabolic responses to high protein diet in Korean elite bodybuilders with high-intensity resistance exercise. Journal of the International Society of Sports Nutrition. (2011); 8(1): 1.

[34] Jacobson BH, Conchola EG, Glass RG, Thompson BJ. Longitudinal morphological and performance profiles for American, NCAA Division I football players. The Journal of Strength & Conditioning Research. (2013); 27(9): 2347–2354.

[35] Norton K, Olds T. Anthropometrica: A Textbook of Body Measurement for Sportsand Health Courses. Australia: University of New South Wales Press; (1996).

[36] Torres-Unda J, Zarrazquin I, Gil J, Ruiz F, Irazusta A, Kortajarena M, Seco J & Irazusta J. Anthropometric, physiological and maturational characteristics in selected elite and non-elite male adolescent basketball players. Journal of Sports Sciences, (2013); 31(2): 196–203.

[37] Gabbett T, Georgieff B. Physiological and anthropometric characteristics of Australian junior national, state, and novice volleyball players. Journal of Strength and Conditioning Research. (2007); 21(3): 902–908. doi:10.1519/00124278-200708000-00042.

[38] Rebelo A, Brito J, Maia J, Coelho-e-Silva MJ, Figueiredo AJ, Bangsbo J, Seabra A. Anthropometric characteristics, physical fitness and technical performance of under-19 soccer players by competitive level and field position. International Journal of Sports Medicine. (2013); 34(04): 312–317.

[39] Cochrane KC, Housh TJ, Smith CM, Hill EC, Jenkins ND, Johnson GO, Cramer JT. Relative contributions of strength, anthropometric, and body composition characteristics to estimated propulsive force in young male swimmers. The Journal of Strength & Conditioning Research. (2015); 29(6): 1473–1479.

[40] Moro VL, Gheller RG, Berneira JDO, Hoefelmann CP, Karasiak FC, Moro ARP, Diefenthaeler F. Comparison of body composition and aerobic and anaerobic performance between competitive cyclists and triathletes. Revista Brasileira de Cineantropometria & Desempenho Humano. (2013); 15(6): 646–655.

[41] Arazi H, Mirzaei B, Nobari H. Anthropometric profile, body composition and somatotyping of national Iranian cross-country runners. Turkish Journal of Sport and Exercise. (2015); 17(2): 35–41.

[42] Vernillo G, Schena F, Berardelli C, Rosa G, Galvani C, Maggioni M, La Torre A. Anthropometric characteristics of top-class Kenyan marathon runners. The Journal of Sports Medicine and Physical Fitness. (2013); 53(4): 403–408.

[43] Belli T, Meireles CLDS, Costa MDO, Ackermann MA, Gobatto CA. Somatotype, body composition and performance in ultramarathon. Revista Brasileira de Cineantropometria & Desempenho Humano. (2016); 18(2): 127–135.

[44] Petroski EL, Fraro JD, Fidelix YL, Silva DAS, Pires-Neto CS, Dourado AC, Viera FS. Anthropometric, morphological and somatotype characteristics of athletes of the Brazilian Men's volleyball team: An 11-year descriptive study. Revista Brasileira de Cineantropometria & Desempenho Humano. (2013); 15(2): 184–192.

[45] Grogan S. Body image and health: Contemporary perspectives. Journal of Health Psychology. (2006); 11(4): 523–530.

[46] Tatangelo G, McCabe M, Mellor D, Mealey A. A systematic review of body dissatisfaction and sociocultural messages related to the body among preschool children. Body Image. (2016); 18: 86–95.

[47] Berge JM, Hanson-Bradley C, Tate A, Neumark-Sztainer D. Do parents or siblings engage in more negative weight-based talk with children and what does it sound like? A mixed-methods study. Body Image. (2016); 18: 27–33.

[48] Brown A. Media use by children younger than 2 years. Pediatrics. (2011); 128(5): 1040–1045.

[49] Harrison S, Rowlinson M, Hill AJ. "No fat friend of mine": Young children's responses to overweight and disability. Body Image. (2016); 18: 65–73.

[50] El Ansari W, Dibba E, Labeeb S, Stock C. Body image concern and its correlates among male and female undergraduate students at Assuit University in Egypt. Global Journal of Health Science. (2014); 6(5): 105–117.

[51] Fiske L, Fallon EA, Blissmer B, Redding CA. Prevalence of body dissatisfaction among United States adults: Review and recommendations for future research. Eating Behaviors. (2014); 15(3): 357–365.

[52] Bruns GL, Carter MM. Ethnic differences in the effects of media on body image: The effects of priming with ethnically different or similar models. Eating Behaviors. (2015): 17: 33–36.

[53] Swami V, Kannan K, Furnham A. Positive body image: Inter-ethnic and rural–urban differences among an indigenous sample from Malaysian Borneo. International Journal of Social Psychiatry. (2012); 58(6): 568–576.

[54] Anderson LM, Reilly EE, Gorrell S, Anderson DA. Running to win or to be thin? An evaluation of body dissatisfaction and eating disorder symptoms among adult runners. Body Image. (2016); 17: 43–47.

[55] Muller SM, Gorrow TR, Schneider SR. Enhancing appearance and sports performance: Are female collegiate athletes behaving more like males? Journal of American College Health. (2009); 57(5): 513–520.

[56] Olvera N, McCarley K, Rodriguez AX, Noor N, Hernández-Valero MA. Body image disturbances and predictors of body dissatisfaction among Hispanic and White preadolescents. Journal of Research on Adolescence. (2015); 25(4): 728–738.

[57] Ferraro F, Muehlenkamp J, Paintner A, Wasson K, Hager T, Hoverson F. Aging, body image, and body shape. Journal of General Psychology. (2008); 135(4): 379–392.

[58] Soenen S, Chapman IM. Body weight, anorexia, and undernutrition in older people. Journal of the American Medical Directors Association. (2013); 14(9): 642–648.

[59] Quick VM, Byrd-Bredbenner C. Disordered eating, socio-cultural media influencers, body image, and psychological factors among a racially/ethnically diverse population of college women. Eating Behaviors. (2014); 15(1): 37–41.

[60] Awad G, Norwood C, Taylor D, Martinez M, McClain S, Jones B, Holman A, Chapman-Hilliard C. Beauty and body image concerns among African American college women. Journal of Black Psychology. (2015); 41(6): 540–564.

[61] Hausenblas HA, Fallon EA. Exercise and body image: A meta-analysis. Psychology and Health. (2006); 21(1): 33–47.

[62] Rodríguez-Camacho DF, Alvis-Gome KM. Overview of the body image and its implications in sport. Revista de la Facultad de Medicina. (2015); 63(2): 279–287.

[63] Hausenblas HA, Downs DS. Comparison of body image between athletes and nonathletes: A meta-analytic review. Journal of Applied Sport Psychology. (2001); 13(3): 323–339.

[64] Ginis KAM, Strong HA, Arent SM, Bray SR, Bassett-Gunter RL. The effects of aerobic-versus strength-training on body image among young women with pre-existing body image concerns. Body Image. (2014); 11(3): 219–227.

[65] Williams PA, Cash TF. Effects of a circuit weight training program on the body images of college students. International Journal of Eating Disorders. (2001); 30(1): 75–82.

[66] Homan KJ, Tylka TL. Appearance-based exercise motivation moderates the relationship between exercise frequency and positive body image. Body Image. (2014); 11(2): 101–108.

[67] Vöcks S, Hechler T, Rohrig S, Legenbauer T. Effects of a physical exercise session on state body image: The influence of pre-experimental body dissatisfaction and concerns about weight and shape. Psychology and Health. (2009); 24(6): 713–728.

[68] Varnes JR, Stellefson ML, Janelle CM, Dorman SM, Dodd V, Miller MD. A systematic review of studies comparing body image concerns among female college athletes and non-athletes, 1997–2012. Body Image. (2013); 10(4): 421–432.

[69] Latorre-Román PÁ, García-Pinillos F, Jiménez-Obra A, Párraga-Montilla J. Exercise dependence and body dissatisfaction in different endurance sports and its relationship with the motivation to sport. Dependencia al ejercicio físico e insatisfacción corporal en diferentes deportes de resistencia y su relación con la motivación al deporte. Revista de psicología del deporte. (2016); 25(1): 113–120.

[70] Goltz FR, Stenzel LM, Schneider CD. Disordered eating behaviors and body image in male athletes. Revista Brasileira de Psiquiatria. (2013); 35(3): 237–242.

[71] DeFeciani L. Eating disorders and body image concerns among male athletes. Clinical Social Work Journal. (2016); 44(1): 114–123.

[72] Rothbart MK. Biological processes in temperament. In Handbook of Temperament, by G. Childhood, A. Kohnstamm, J. Bates, and M. K. Rothbart (Eds.), 1989, Sussex, England: Wiley; (1989), pp. 77–110.

[73] Kagan J, Snidman N, Arcus D, Reznick JS. Galen's Prophecy: Temperament in Human Nature. New York: Basic Books Galen's prophecy; (1994). XXIV, 376 p.

[74] Wiltse LL, Pait TG. Herophilus of Alexandria (325-255 BC): the father of anatomy. Spine. (1998); 23(17): 1904–1914.

[75] Harcourt G. Andreas Vesalius and the anatomy of antique sculpture. Representations. (1987); (17): 28–61.

[76] Fosbøl MØ, Zerahn B. Contemporary methods of body composition measurement. Clinical Physiology and Functional Imaging. (2015); 35(2): 81–97.

[77] Wang Z, Wang ZM, Heymsfield SB. History of the study of human body composition: A brief review. American Journal of Human Biology. (1999); 11(2): 157–165.

[78] Neamat-Allah J, Johnson T, Nabers D, Hüsing A, Teucher B, Katzke V, Kühn T. Can the use of blood-based biomarkers in addition to anthropometric indices substantially improve the prediction of visceral fat volume as measured by magnetic resonance imaging? European Journal of Nutrition. (2015); 54(5): 701–708.

[79] Mazić S, Lazović B, Đelić M, Suzić-Lazić J, Aćimović T, Brkić P. Body composition assessment in athletes: A systematic review. Medicinski pregled. (2014); 67(7–8): 255–260.

[80] Knight GS, Beddoe AH, Streat SJ, Hill GL. Body composition of two human cadavers by neutron activation and chemical analysis. American Journal of Physiology-Endocrinology and Metabolism. (1986); 250(2): E179–E185.

[81] Martin AD, Ross WD, Drinkwater DT, Clarys JP. Prediction of body fat by skinfold caliper: Assumptions and cadaver evidence. International Journal of Obesity. (1984); 9: 31–39.

[82] Mitsiopoulos N, Baumgartner RN, Heymsfield SB, Lyons W, Gallagher D, Ross R. Cadaver validation of skeletal muscle measurement by magnetic resonance imaging and computerized tomography. Journal of Applied Physiology. (1998); 85(1): 115–122.

[83] Bazzocchi A, Diano D, Ponti F, Salizzoni E, Albisinni U, Marchesini G, Battista G. A 360-degree overview of body composition in healthy people: Relationships among anthropometry, ultrasonography, and dual-energy x-ray absorptiometry. Nutrition. (2014); 30 (6): 696–701.

[84] Popuri K, Cobzas D, Esfandiari N, Baracos V, Jägersand M. Body composition assessment in axial CT images using FEM-based automatic segmentation of skeletal muscle. IEEE Transactions on Medical Imaging. (2016); 35(2): 512–520.

[85] Dall'Ara E, Eastell R, Viceconti M, Pahr D, Yang L. Experimental validation of DXA-based finite element models for prediction of femoral strength. Journal of the Mechanical Behavior of Biomedical Materials. (2016); 63: 17–25.

[86] Goosey-Tolfrey V, Keil M, Brooke-Wavell K, De Groot S. A comparison of methods for the estimation of body composition in highly trained wheelchair games players. International Journal of Sports Medicine. (2016); 37(10): 799-806. doi:10.1055/s-0042-104061

[87] Wagner DR, Cain DL, Clark NW. Validity and reliability of A-mode ultrasound for body composition assessment of NCAA division I athletes. PloS One. (2016); 11(4): e0153146.

[88] Thomas EL, Saeed N, Hajnal JV, Brynes A, Goldstone AP, Frost G, Bell JD. Magnetic resonance imaging of total body fat. Journal of Applied Physiology. (1998); 85(5): 1778–1785.

[89] González-Neira M, San Mauro-Martín I, García-Angulo B, Fajardo D, Garicano-Vilar E. Nutritional assessment, evaluation of body composition and its relationship with athletic performance in a women's soccer team. Valoración nutricional, evaluación de la composición corporal y su relación con el rendimiento deportivo en un equipo de fútbol femenino. Revista española de nutrición humana y dietética. (2015); 19(1): 36–48.

[90] Marinho BF, Andreato LV, Follmer B, Franchini E. Comparison of body composition and physical fitness in elite and non-elite Brazilian jiu-jitsu athletes. Science & Sports. (2016); 31(3): 129–134.

[91] Guiraudou M, Maimoun L, Dumas JM, Julia M, Raingeard I, Brun JF. Composition corporelle mesurée par impédancemétrie segmentaire (BIAS) et performance de sprint chez les rugbymen. Science & Sports. (2015); 30(5): 298–302.

[92] Franchini E, Brito CJ, Artioli GG. Weight loss in combat sports: physiological, psychological and performance effects. Journal of the International Society of Sports Nutrition. (2012); 9(1): 1.

[93] Khodaee M, Olewinski L, Shadgan B, Kiningham, RR. Rapid weight loss in sports with weight classes. Current Sports Medicine Reports. (2015); 14(6): 435–441.

[94] Kurt C, Sagiroglu I. Rapid weight loss practice and perceived problems during reduction periods of Turkish young combat athletes. Journal of Physical Education and Sport. (2015); 15(4): 748.

[95] da Silva PRP, de Souza Trindade R, De Rose EH. Composição corporal, somatotipo e proporcionalidade de culturistas de elite do Brasil. (2003); **Rev Bras Med Esporte**, 9: 403–407.

[96] Mielgo-Ayuso J, Calleja-González J, Clemente-Suárez VJ, Zourdos MC. Influence of anthropometric profile on physical performance in elite female volleyballers in relation to playing position. Nutr Hosp. (2014); 31(2): 849–857.

[97] Kenney WL, Wilmore J, Costill D. Physiology of Sport and Exercise 6th Edition. USA: Human Kinetics; (2015). 626 p.

[98] Arnold EM, Hamner SR, Seth A, Millard M, Delp SL. How muscle fiber lengths and velocities affect muscle force generation as humans walk and run at different speeds. Journal of Experimental Biology. (2013); 216(11): 2150–2160.

[99] Mangine GT, Fukuda DH, LaMonica MB, Gonzalez AM, Wells AJ, Townsend JR, Hoffman JR. Influence of gender and muscle architecture asymmetry on jump and sprint performance. Journal of Sports Science & Medicine. (2014); 13(4): 904.

[100] Terzis G, Spengos K, Kavouras S, Manta P, Georgiadis G. Muscle fibre type composition and body composition in hammer throwers. Journal of Sports Science & Medicine. (2010); 9(1): 104–109.

Chapter 3

Activity Tracking and Improved Health Outcomes

Daniel V. Gaz, Thomas M. Rieck and
Nolan W. Peterson

Additional information is available at the end of the chapter

http://dx.doi.org/10.5772/65240

Abstract

Activity tracking devices are a popular way for health-minded individuals to measure daily movement and estimate energy expenditure. Tracking, in its many forms, has been proven to improve health outcomes. From paper diaries to devices and smartphone applications, these data are becoming more important and have the potential to be used in the physician's office as part of a realistic physical activity plan for improved health outcomes. As a health-care professional, amid a rapidly expanding accelerometer market, it is important to know the application and practicality of these devices, as well as the evidence behind an individual's usage. What we choose to recommend, as health-care professionals, can influence the health and well-being of patients in the wellness arena. This chapter will look at research focused on activity tracking, including metrics such as sleep, nutrition, and physical activity, and health outcomes to further educate health-care professionals in an ever-evolving field.

Keywords: activity tracking, physical activity, health, wellness, behavior modification, accelerometer, disease prevention

1. Introduction

Man has been physically active since our days as primitive nomads [1]. From the Paleolithic Era, where we hunted and gathered to stay alive, to the days of the ancient Greeks and Romans, who placed a high emphasis on exercise and fitness, we have typically necessitated and revered physical activity. Recent research [2] has begun to take a look at the decline in our daily energy expenditure and, since the 1950s, the results have been shocking. In the United States alone, there has been a significant increase in sedentary time, coupled with a decrease in folks employed in careers that have a high level of physical activity [2]. Over the last 50 years, it has been estimated that daily occupation-related energy expenditure has decreased by more than

100 calories, accounting for a significant portion of the increase in mean body weights for women and men in the United States [3]. Many people remain sedentary for a time that equals a normal working shift—close to 7.7 h/day—according to a recent study [4] that looked at 6329 participants and the level of activity they achieved during a 10-h work day. With such a dramatic change in the amount of regular physical activity that our society accrues on a daily basis, it is important to understand the negative implications of the direction we were headed. Regular physical activity provides us with myriad of health benefits [5]. These include, but are not limited to, weight control [6], health condition and disease prevention [7], improvements in mood [8], decreased fatigue [9], and better sleep quality [10].

Even though the benefits from exercise are known, a large population of people does not meet the minimum physical activity guidelines [11], as defined by the American College of Sports Medicine (ACSM). The ACSM physical activity recommendations include "moderate-intensity aerobic (endurance) physical activity for a minimum of 30 minutes on five days each week or vigorous-intensity aerobic physical activity for a minimum of 20 minutes on three days each week" [11]. Accruing 150 min of physical activity at a moderate level (65% $V^{\bullet}O_{2max}$) or 60 min of vigorous activity (80% $V^{\bullet}O_{2max}$), as defined by ACSM [12], does not seem like an arduous task, yet so many people fail to meet these minimum requirements. According to the Centers for Disease Control and Prevention (CDC), only 20% of adults (roughly 50 million men and women as of 2013) meet the overall physical activity recommendations [13]. This lack of overall fitness and physical activity has been shown to contribute directly to an increase in mortality risk [14], among other diseases and conditions. Researchers have even compared it to smoking, with one in 10 deaths being attributed to a sedentary lifestyle [15].

Buried among this doom and gloom of our sedentary society lie the health benefits for those afflicted with acute and chronic diseases, such as diabetes, obesity, cardiovascular disease (CVD), cancer, and other conditions that can be positively affected with regular physical activity. Whether it is a complete elimination of signs and symptoms, or a mitigation of certain chronic disease profiles, physical activity has shown promise in improving quality of life for those with modifiable and nonmodifiable diagnoses [16]. A recent meta-analysis [17] found that individuals who were physically active at levels lower than the minimum amount recommended (150 min of moderate activity per week) also had a significantly lower risk of coronary heart disease (CHD). This is an important finding, as those with sedentary lifestyles and, as a result, low levels of fitness, can reduce their risk for disease with very modest and achievable levels of physical activity. Two Harvard studies showed how little exercise is necessary to reduce the risk of disease. In one paper, walking just ten blocks a day lowered heart disease risk by 33% [18], and in another paper, walking just one hour per week was enough to reduce heart disease risk by 20% [19]. Even in those who are overweight and/or obese, but still exercise regularly, the outcomes are better than individuals who are thin and fail to meet daily exercise recommendations. "What we're learning is that a body that exercises regularly is generally a healthy body, whether that body is fat or thin," says Glenn Gaesser, PhD, a professor of exercise and wellness at Arizona State University [20]. "The message should really be that if you are exercising regularly, you shouldn't necessarily be looking at the scale to determine how healthy or fit you are," Gaesser says [20].

http://dx.doi.org/10.5772/65240

One very popular way to keep track of accumulated physical activity is body-mounted accelerometers. Early versions were simplistic in nature. In 1780, Swiss horologist Abraham-Louis Perrelet created the first pedometer that measured both steps and distance while walking, basing this ingenious invention on a 1770 mechanism that powered a self-winding watch [21]. Since Perrelet's novel development, the field of accelerometry has grown into a projected $5.9 billion industry by 2019 [22]. Brands like Garmin, Jawbone, and Fitbit have captured the activity tracking market, while Samsung and Apple take hold of the smartwatch domain. In between lay thousands of smartphone applications that you can download to turn your phone into a de facto activity tracker, ranging from free to $1.99. According to a recent (2015) publication in BMC Medicine, "consumer wearable devices for activity tracking have shown promise in post-surgery recovery in cardiac patients, pulmonary rehabilitation, and activity counseling in diabetic patients, among others" [23]. An earlier study (2007) suggested that "the use of a pedometer is associated with significant increases in physical activity and significant decreases in body mass index and blood pressure" [24]. Specifically, subjects that tracked their steps, but did not have a step goal, averaged 686 more steps per day. Subjects that were given a 10,000 step/day goal increased their overall step count by 2998 steps. Subjects that were given a step goal, and also gradually increased their steps over the course of the study, got 3175 more steps and an additional 1200 min of "non-seated" time. Overall, pedometer wearers saw a rise in physical activity of 27% [24].

This chapter, appropriately titled "Activity Tracking and Improved Health Outcomes," will go into further detail on the benefits of using devices like pedometers and smartwatches to track physical activity and exercise, while also investigating the improvement of health outcomes as a result. Research in this area is gaining momentum and traction, but the data is not a slam-dunk. Some studies show good correlations between tracking activity and improving overall health [25], while others still believe that more work needs to be done, as it is a challenge to track accurately and use the information advantageously [26]. Regardless, clinicians and researchers have come to one conclusion—the widespread availability to collect patient-centered information on their movement, nutrition, exercise, and sleep can modify risk factors, further concrete healthy behavior change, and look at other outcomes like postsurgical activity or changes in daily lifestyle habits when given new medications or dosages [27]. Our society is slowly, but surely, becoming more and more quantified [28]. There is no better time than now to understand how this quantified self can improve health outcomes with minimal financial and time investment.

2. The health benefits of exercise

Our society has long been aware of the benefits and importance of regular physical activity. For much of our young lives, it was a mandatory part of our regular education [29, 30], and a significant contributor to public health [31, 32]. Sallis noted that "for maximal public health benefit, school physical education programs should prepare children for a lifetime of physical activity" [32]. It makes sense that schools would devote class time to increase the amount of physical activity, given the intended consequences of more health and wellness down the line.

As school children, we were exposed to hours of recess and physical activity, and summers spent playing outdoors with friends and family. For some of us, as we progressed to middle and high school, play turned to sport. Physical activity was now a component of achieving success on the court or playing field. For those who did not compete in a sport, physical education class might have been our only option for an increase in caloric expenditure during the school day. It is at this point when we begin contemplating purposeful exercise—something is more commonly known as "working out." According to Dr. David Bassett, Jr., a professor in the Department of Exercise, Sport, and Leisure Studies at the University of Tennessee, Knoxville, exercise is a specific form of physical activity that is planned, purposeful, and performed with the intention of acquiring fitness or other health benefits. Examples include working out at a health club or recreation center, swimming, cycling, running, and recreational sports [33].

As Dr. Bassett mentioned, purposeful exercise has an inherent intention to acquire fitness, along with other health benefits. These benefits, long researched, in the form of recommendations [34] and position stances [35–39] published regularly to increase the awareness of regular physical activity and the importance of such. The most recent position stance on exercise, courtesy of the ACSM, recommends "a comprehensive program of exercise including cardiorespiratory, resistance, flexibility, and neuromotor exercise of sufficient volume and quality…for apparently healthy adults of all ages" [39]. This stance, focused primarily on healthy adults, delves into specific benefits of regular exercise that include a lower risk of all-cause mortality and cardiovascular disease with increased levels of cardiorespiratory fitness [39–43], a risk reduction in developing coronary heart disease (CHD), stroke, type 2 diabetes (T2D), and some forms of cancer (e.g., colon and breast cancers) [44]. Additionally, regular exercise and physical activity lower blood pressure, have been shown to improve lipoprotein profile, C-reactive protein, and other CHD biomarkers, enhance insulin sensitivity, and play an important role in weight management [44].

Clearly, evidence supports the health benefits of exercise in adults, especially when performed on a regular basis. While exercise remains a somewhat ambiguous term or phrase, it is clear that a well-rounded program, consisting of cardiovascular and resistance training, as well as regular flexibility and neuromotor (read—functional fitness) exercise gives an individual an excellent, and beneficial, chance to improve their overall health profile. More specifically, the literature supports the following recommendations:

"The ACSM recommends that most adults engage in moderate-intensity cardiorespiratory exercise training for ≥30 min d^{-1} on ≥5 d wk^{-1} for a total of ≥150 min wk^{-1}, vigorous-intensity cardiorespiratory exercise training for ≥20 min d^{-1} on ≥3 d wk^{-1} (≥75 min wk^{-1}), or a combination of moderate- and vigorous-intensity exercise to achieve a total energy expenditure of ≥500–1000 MET min wk^{-1}. On 2–3 d wk^{-1}, adults should also perform resistance exercises for each of the major muscle groups, and neuromotor exercise involving balance, agility, and coordination. Crucial to maintaining joint range of movement, completing a series of flexibility exercises for each of the major muscle-tendon groups (a total of 60 s per exercise) on ≥2 d wk^{-1} is recommended [39].

While it would be easy to give blanket physical activity recommendations for children and elderly adults, the above-mentioned recommendations cannot be directly translated to the youth and the older population. Still—both groups benefit greatly from regular exercise with comparable, and just as important, improvements in overall health and wellness. Children do not consider purposeful exercise. Rather, they look at opportunities to play games and sports with friends and family, both at school and at home, as ways to have fun and socialize. When children engage in regular physical activity, they are less likely to become obese [45], will have stronger muscles [46] and bones [47], lower their risk of developing type 2 diabetes [48], and have a better outlook on life [49]. They will also sleep better [50] and learn to handle physical and emotional challenges better [51]—skills that will undoubtedly translate when they reach their teenage and adult years. The Centers for Disease Control and Prevention (CDC) recommend children and adolescents accumulate 60 or more minutes of activity each day. Most of those 60 min should be moderate (walking) to vigorous (running) aerobic activity, in addition to 3 days of strength and resistance exercises (gymnastics or push-ups) per week, and 3 days a week of bone-strengthening activities (jumping rope or running) [52]. The World Health Organization (WHO) recommends children and youth (5–17 years old) accumulate at least 60 min of moderate- to vigorous-intensity physical activity daily, with amounts of physical activity greater than 60 min providing additional health benefits [53]. WHO and CDC concur on the bulk of a child's daily activity being aerobic, incorporating vigorous-intensity activities should, including those that strengthen muscle and bone at a minimum of three times per week [52, 53].

As we age, we become progressively more sedentary. The reasons behind this are multifactorial —elderly have self-reported barriers to exercise including inertia, fear of falling/safety, time, negative affect, physical ailments, social, discomfort, weather, age, inconvenience, perceived capability, and verbal persuasion [54]. With a rapid increase in the population of men and women over the age of 65, it is important to know about how to overcome the barriers to exercise that the elderly deal with on a regular basis. Specific recommendations include identifying realistic goals, such as making it easier for an older adult to play with loved ones, improving performance in activities of daily living, and improving the overall quality of life. These strategies might improve exercise adherence and eliminate the above-mentioned perceived barriers [54]. It can be difficult to persuade the elderly to become physically active. As self-reported above, the elderly often believe themselves to be too old or frail for physical activity.

When compared to prescriptions for better health, exercise is rarely viewed as a surrogate to medication. Older adults also encounter more barriers to physical activity and exercise participation [55]. Even so, exercise recommendations for the elderly exist. To promote and maintain health, recommendations include moderate-intensity aerobic exercise 5 days a week for a minimum of 30 min, or vigorous intensity aerobic activity 3 days a week for a minimum of 20 min [56]. Strength and resistance training is also recommended for older adults. Recommendations include performing activities that maintain or increase muscular strength and endurance for a minimum of 2 days each week [56]. Examples include push-ups, squats, biceps curls, triceps extensions, and any core or abdominal exercises. Ideally, one should perform 8–

10 exercises, using major muscle groups, on two or more nonconsecutive days per week. Flexibility and balance are also excellent additions to a well-rounded exercise program for the elderly. Older adults should regularly perform activities with a goal of maintaining or increasing flexibility at least twice a week for a minimum of 10 min each day. Additionally, older adults that are independent and live without assistance should perform exercises that maintain or improve balance to decrease the substantial risk of falls [56, 57]. Since more than 1.3 million adults in the U.S. visit the emergency department for fall-related injuries each year, exercise can have a drastic impact on fall prevention [58]. It is important for folks in their sixth decade and beyond to exercise, as it can reduce the risk of falls and injuries from falls [59], prevent functional limitations [60–63], and can serve as effective therapy for many chronic diseases, many of which will be covered in the next section.

3. Disease prevention and mitigation with exercise

As covered in the previous section, there is a clear relationship [64, 65] between improved health and wellness and the inclusion of regular physical activity. While it would be an ideal scenario for everyone to accumulate the recommended bouts of exercise week after week, research has shown that we are less active on a regular basis, especially in the last 50 years [3]. With an increase in sedentary time comes an increase in diseases [66] that could, conceivably, be prevented with more movement throughout the day. Over the next few pages, ways to prevent, mitigate, or eliminate chronic or modifiable diseases, like diabetes and hypertension (HTN), will be discussed. The focal point will be on specific exercise recommendations and outcomes that revolve around incorporating physical activity with the sole goal of disease prevention and elimination.

When one thinks of diseases that are related to inactivity and a sedentary lifestyle, many come to mind. According to the WHO, "sedentary lifestyles increase all causes of mortality, double the risk of cardiovascular diseases, diabetes, and obesity, and increase the risks of colon cancer, high blood pressure, osteoporosis, lipid disorders, depression and anxiety" [67]. There is another conformation of diseases that result from a lack of daily caloric expenditure called metabolic syndrome. It is defined as having three of the following five conditions—abdominal obesity, high triglycerides, low high density lipoprotein (HDL) cholesterol, high blood pressure, and high fasting blood sugar [68]. Regardless of what disease or diseases one has, it has been proven that exercise can help prevent, reduce, or eliminate the signs and symptoms of one or more acute or chronic conditions.

3.1. Exercise and mortality risk

When we think of exercise, one thing that comes to mind is how it can positively effect and improve our quality of life. We rarely think of how it can prevent us from dying early, yet data supports this "fact of life." Fortunately, it does not take much to move the needle. Those who did a little movement (less than the amount outlined in the 2008 Physical Activity Guidelines for Americans) still had a 20% lower mortality risk when compared with individuals who did no exercise at all. Those who exercised for the minimum recommended physical-activity target had a 31% lower risk of dying compared with their sedentary counterparts. Those who did a lot more, including those who exceeded the weekly recommendations, had an even larger reduction in mortality risk. On the extreme ends of the exercise spectrum, individuals who performed 3–5 times the recommended minimum for physical activity had nearly a 40% mortality risk reduction when compared to sedentary individuals [69]. It is clear that an exercise dose-response exists and, as this data shows, it seems to be a minimal dose with a maximal response when looking at mortality risk. Studies have taken a deeper look into specific exercise modalities—one being a very popular and widespread form of aerobic exercise: jogging. In its Outdoor Participation Report, the Outdoor Foundation noted that 31.6 million adults over the age of 25 went running or jogging in 2016 [70]. Such a widespread physical activity can have beneficial effects on health and wellness, even in small doses. One research study that compared joggers to nonjoggers, completing as little as 1–2.4 h/week of jogging, had lower mortality rates. Based on these results, the authors recommend modest jogging recommendations of 2–3 times per week for a total cumulative duration of 2.5 h [71]. While it seems counterintuitive, it does not take much physical activity to improve our chances at survival and longevity. If nothing else, some light cardiovascular activity can improve the time we spend alive, with friends and family. Since the exercise recommendations are modest, it is important to spread the word to those who need help getting started on a regular activity program.

3.2. Exercise and cardiovascular disease risk

Cardiovascular disease (CVD) is prevalent across both genders and all backgrounds and nationalities. According to a recent report from the American Heart Association [72], one of every three deaths in the U.S. in 2013 was from heart disease, stroke, and other CVDs. Additionally, heart disease and stroke were the top two killers worldwide. In the United States, CVDs claimed 801,000 lives, and heart disease killed more than 370,000 people. 795,000 people suffered strokes, 129,000 of which did not survive. 116,000 of the 750,000 people in the U.S. who had a heart attack died [72]. Even cost is an issue with those afflicted with CVD—costs for each U.S. state can be up to $26 billion, and the cost of lost work days can be up to $1.3 billion [73]. Damning evidence, but what can we do about it? Is exercise the answer to combat the risk of CVD? As we learned above, it can help improve our quality of life and decrease the risk of mortality. As CVD claims nearly a million lives in the United States each year, it would be presumptive that exercise could work in a similar fashion.

It has been repeatedly shown that an inverse relationship exists between physical activity and the occurrence of CVDs. This means that with an increase in physical activity, the relative risk of developing CVD is decreased [74]. Regular physical activity using large muscle groups (quadriceps, hamstrings, as well as the muscles of the chest, back, and upper arms), like walking, running, or swimming, produces cardiovascular adaptations that increase exercise capacity, endurance, and skeletal muscle strength. Routine physical activity also prevents the development of coronary artery disease (CAD) and reduces symptoms in patients with established cardiovascular disease [75]. Thus, it has been proven that we can reduce the likelihood of CVD with regular exercise. How much mitigation or prevention remains to be seen, but we know through numerous studies that there is a positive relationship that exists.

3.3. Exercise and diabetes prevention/mitigation

Since 1980, the number of people with diabetes has risen from 108 million to 422 million in 2014, and the global prevalence of diabetes among adults over the age of 18 has risen from 4.7% to 8.5% in 2014 [76]. In areas of middle and low-income, diabetes prevalence is trending upwards, and contributes to blindness, kidney failure, heart attacks, stroke, and lower limb amputation [76]. 1.5 million deaths were caused by diabetes and another 2.2 million deaths were attributed to high blood glucose in 2012 [76], with half of those deaths attributable to high blood glucose occurring before 70 years of age [76]. The World Health Organization projects that diabetes will be the 7th leading cause of death by 2030 [77], a staggering thought considering the relative stranglehold that cancer and heart disease have on mortality.

A great deal of evidence has been collected that supports physical activity, among other therapies, may be useful in preventing or delaying the onset of type 2 diabetes (T2D). There are now three published trials documenting that, with lifestyle modification (weight loss and regular moderate physical activity), diabetes can be delayed or prevented [78–81]. Additionally, other research has proven that long-term interventions, using both diet and exercise, leads to substantial metabolic improvement that may contribute to prevent or postpone manifest diabetes [82]. In the case of T2D, exercise is truly medicine. Increasing physical activity can reduce the incidence of T2D by almost 60% in people at risk [83].

3.4. Exercise and obesity

While increased caloric intake is often blamed for rising rates of obesity, no association between these two variables was found in a recent study, which looked at data from the National Health and Nutrition Examination Survey (NHANES) from 1988 to 2010. More importantly, and in the interest of this particular chapter, an association was found between the trends over time for lack of physical activity and high BMI numbers [84]. Society is less active than we once were, especially at work, where we expend roughly 100–150 calories/day less than what we used to in the 1960s. Researchers from this study noted that such a dramatic reduction in daily calorie burn from our jobs alone, extrapolated over the course of a year, could explain the steady rise in weight gain [3]. This weight gain is not limited to adults alone. A recent study found that roughly a third of obese preschool children were obese as adults, and about half of obese school-age children were obese as adults. Across all conditions, the risk of adult obesity was at least twice as high for obese children as for non-obese children. The risk of adult obesity was greater for children who were at higher levels of obesity and for children who were obese at older ages [85]. Children or adults alike, when individuals increase physical activity, there is a positive relationship with reductions in total adiposity [86]. It has also been shown that physical activity is associated with reduction in abdominal and visceral fat, both of which can contribute to other diseases within the body, particularly the cardiovascular system [87, 88]. More specifically, the more exercise and activity one gets, the less fat an individual has. We

http://dx.doi.org/10.5772/65240

need to be physically active to burn more calories and, in turn, see reductions in body composition.

3.5. Exercise and cancer prevention

Cancer and exercise have a tricky relationship. Often, those diagnosed and undergoing treatment for cancer do not have the energy for physical activity, yet it can be beneficial for one's overall health and well-being, especially in the throes of such a debilitating disease. Women undergoing chemotherapy or radiation therapy for breast cancer that maintained a regular exercise routine (at least 90 min in duration per week for 3 or more days) felt less fatigued and emotionally distressed, and had higher functional ability and quality of life when compared to their less active counterparts [9]. A recent meta-analysis found an inverse association between physical activity and colon cancer. Individuals can likely reduce their risk of colon cancer with the addition of physical activity [89]. Breast cancer seems to be just as responsive, as there is reasonably clear evidence that physically active women have about nearly a third of the risk when compared to inactive women. Even more, just 30–60 min of moderate to vigorous activity per week has been shown to reduce the risk of breast cancer [89]. Like the previously mentioned mortality rates, it does not take much to reduce the likelihood of such a terrible disease like cancer. In addition to risk reduction, there is also the benefit of exercise and cancer survivorship. Participants with higher levels of physical activity following a cancer diagnosis were less likely to have a cancer recurrence and had increased survival [90]. Over 20 studies have examined the impact of physical activity on the risk of lung cancer. Overall, this wealth of data suggests an inverse association between physical activity and lung cancer risk. In fact, the most physically active individuals experienced about a 20% reduction in risk. A theme is recurring—exercise is medicine [91].

3.6. Exercise and hypertension

Exercise is advocated for the prevention, treatment, and control of high blood pressure or hypertension (HTN) [92]. When left uncontrolled, high blood pressure can lead to stroke by damaging and weakening blood vessels in the brain, causing them to narrow, rupture, or leak. High blood pressure can also cause blood clots to form in the arteries leading to the brain, blocking blood flow, and potentially causing a stroke. It can also cause arterial and kidney complications, in addition to the eyes, bones, and sleep [93]. Exercise intervention studies have shown that regular aerobic exercise significantly lowers blood pressure in hypertensive individuals. Regular physical activity—somewhere in the ballpark of 30 min on most days of the week—can lower the blood pressure by 4–9 mm of mercury [94]. As exercise intensity moves from moderate to vigorous, it may be even more effective in lowering blood pressure. Since exercise-induced reductions in resting blood pressure and prevention of abnormal increases in blood pressure during physical exertion can lead to fewer cardiovascular events, it makes sense that the same benefits for exercise and CVD are correlated to HTN. Most importantly, they can allow folks with HTN to come off medications, see fewer medication-related side-effects, and, as has been repeated, improve quality of life [95]. High blood pressure is not a terminal diagnosis, but it can lead to further complications that can be quite deleterious if left uncontrolled. It is important to begin stressing exercise as a modifier in order to eliminate the stress that HTN places on the cardiovascular system.

3.7. Exercise and osteoporosis

Often thought of as a disease that only impacts older women, osteoporosis is characterized by low bone mineral density (BMD), coupled with deterioration of bone tissue. These two factors contribute to an increase in bone fragility, leaving those afflicted with this systemic skeletal disease more susceptible to fracture. While fractures might not seem like a big deal, as we age, it becomes more and more important to avoid them at all costs. Nationally, an estimated 13–18% of women and 1–4% of men age 50 and older have osteoporosis; an additional 37–50% of women and 28–47% of men age 50 and older have low bone mass [96]. A recent meta-analysis revealed that women sustaining a hip fracture were five times as likely and men nearly eight times as likely to die within the first 3 months as compared with age- and sex-matched controls [97]. Given the fragility of the skeletal system for those with osteoporosis, it is important to know how to prevent fractures and preserve bone strength.

Osteoporosis is a silent disease that does not cause pain or outward signs of an underlying cause or condition. It has become the most common bone disease worldwide, thus making it a major health problem. One of the contributing factors to maintaining bone health is regular weight-bearing exercise throughout the life span. Moderate exercise protects against osteoporosis, but too little or extremely prolonged bouts of exercise may cause osteoporosis [98]. Unfortunately, as we age, we progressively become sedentary [99]. This is largely attributed to a lack of energy, muscle mass, adequate caloric intake, and fear of injuring oneself and becoming less independent. Disuse and inactivity can cause bone loss, whereas weight-bearing exercises may maintain or improve bone mineral density. There is a significant correlation between muscle strength and bone mineral density, as well as enough evidence to support strengthening exercises leading to an increase in BMD. A well-rounded exercise program, incorporating cardiovascular and resistance training, as well as balance exercises, may prove to be effective for retarding age-related bone loss [100]. At the very least, with more weight-bearing movement (e.g., jogging, running, hopping, or skipping), the likelihood of physical inactivity-related damage to the skeletal system is reduced.

3.8. Exercise and lipid disorders

Physical activity, which involves working major muscle groups like the quadriceps and hamstrings, is vital when controlling lipid metabolism and preventing lipid disorders. Increased physical activity induces a number of positive changes in the metabolism of lipoproteins—serum triglycerides are lowered and the high density lipoprotein (HDL) production is increased [101]. Intervention programs that revolve around daily physical activity can have a positive effect on the concentrations of plasma lipids and lipoproteins, thus reducing the risk of coronary artery disease (CAD) [102]. Specifically, programs that expend roughly 1200–2200 kilocalories of energy per week (kcal/week), the equivalent of 15–20 miles/week of brisk walking or jogging, is associated with significant triglyceride reduction and HDL increases. It can be posited that, for most individuals, the benefits of regular physical activity are present at low training volumes, so long as total caloric expenditure is between 1200 and 2200 and accrue kcals/week. Weekly caloric expenditures through purposeful exercise that meet or exceed the higher end of this range are more likely to produce the desired lipid changes. As far as lipid control is concerned, it is clear that more exercise is better. So much is already known about the benefits of exercise and the concomitant lipid and lipoprotein modifications, as well as the physiological mechanisms for such changes. Exercise professionals and those working in preventative medicine can strive to develop more comprehensive exercise programming to optimize specific caloric expenditure to take advantage of such personalized data and knowledge [103].

3.9. Exercise and depression

Depression is a mood disorder that causes a persistent feeling of sadness and loss of interest. It can affect how you feel, think, and behave and can lead to a variety of emotional and physical problems, including issues with activities of daily living, or even feeling as if life is not worth living [104]. Fortunately, exercise has been shown to benefit those who suffer from depression. A study that looked at individuals with major depression found that those who participated in 4 months of aerobic exercise saw a significant reduction in their depression symptoms. Even more powerful was that if they kept exercising, there was a significant therapeutic benefit over time, with regards to a relapse in symptoms [105]. In short, if they kept up with an exercise program, the antidepressant benefits continued. A review of the literature suggests that exercise is more effective in treating depression than no treatment, and is as effective as psychotherapy and antidepressant medication [106, 107]. Unfortunately, the majority of studies has significant methodological shortcomings. These conclusions are consistent with three recent meta-analyses [108]. In the most recent meta-analysis, Lawlor and Hopker [108] found that exercise was associated with a greater reduction in depressive symptoms when compared with no treatment, and was as effective as cognitive therapy. However, because of the poor quality of evidence reviewed, the effectiveness of exercise in reducing symptoms of depression could not be determined. Therefore, exercise professionals and preventative medicine specialists should proceed with caution when looking for alternative means to help patients afflicted with a depression diagnosis.

3.10. Exercise and anxiety

Anxiety is a somewhat ubiquitous term. The most common form of anxiety, general anxiety disorder (GAD), has a wide range of symptoms that frequently vary. They include, but are not limited to persistent worrying or obsession, inability to set aside or let go, inability to relax, difficulty concentrating, worrying, and distress about decision making [109]. The relationship between exercise and anxiety has been extensively examined over the last 15 years [110], and the results are somewhat similar to its common counterpart, depression, reviewed above. According to the Anxiety and Depression Association of America, it is not uncommon for someone with an anxiety disorder to also suffer from depression or vice versa. Nearly one-half of those diagnosed with depression are also diagnosed with an anxiety disorder [111]. Individuals taking part in exercising conditions significantly reduced anxiety, and additional analyses indicated that subjects experienced an acute increase in anxiety during exercise that subsided upon completion of the exercise bout [112]. More specifically, men who ran on a treadmill at 75% of their VO_{2max} for 30 min saw significant reductions in anxiety following exercise at 10-min intervals up to 30 min [113]. In a different study, subjects exercised at a lesser intensity (70% of VO_{2max}) saw a similar reduction in anxiety, when compared to meditation and quiet rest [114]. Other research has found that aerobic exercise of at least 21 min seems necessary to achieve reductions in anxiety, thus offering therapeutic benefits for reducing anxiety without the dangers or costs of drug therapy or psychotherapy [110]. Perhaps it might be time to pursue nonpharmacological options when treating patients with GAD. The data seems to strongly support such a notion.

3.11. Disease mitigation and prevention conclusion

With all the evidence so strongly supporting physical activity as a way to modify certain lifestyle and health issues, it seems as though more practitioners would prescribe exercise and employ the help of trained exercise science professionals. Unfortunately, health care does not cover the vast majority of services that can be performed by health and wellness specialists, who have an extensive background in disease prevention through exercise, as well as exercise prescription and modification for special populations. Nevertheless, books such as this will serve as the vanguard for providers looking to add more than just a new prescription to their patient's medicine cabinet. It is much easier to engage in physical activity, in the form of walking, cycling, or resistance training exercises when you know the cumulative benefits and how they can alter the potential risks of a sedentary lifestyle. It is not a short-term fix—it is a lifestyle change that has been shown to benefit individuals in need of a change to their quality of life.

4. Tracking and disease prevention/mitigation

As simple as it seems, many of the diseases and conditions mentioned in the previous section have been shown to either resolve or reduce in severity when those afflicted undergo an exercise program with regular physical activity. The research documenting the health benefits of exercise is convincing, but finding ways to encourage sedentary patients to become more active remains a challenge [115]. One concept that is relatively new is keeping track of total exercise minutes from week to week, in an attempt to accumulate the recommended minutes of exercise to see the specific benefits related to the disease(s). Unfortunately, human beings are not as good as remembering exercise times, or many other things, as we might think. Thus, the importance of tracking health data and information is at a premium in our multitasking society. According to Pew Research [116], 70% of the US adults track at least one health metric daily, but the array of how they track varied. 43% were "head trackers," meaning they kept track in their heads—essentially memorizing what they did; 41% used more traditional methods of writing it down on paper; and 23% used technology to track health data and statistics [116]. Given the prevalence of commercially available activity tracking devices, like Fitbit, we can overcome our forgetful nature as human beings and rely on technology to count our steps and activity. No matter what method individuals used to keep track of their data, 40% stated that tracking has changed their health approach [116]. We all keep lists, whether they are for groceries, errands, or job-related tasks that need to be accomplished by the end of the day, week, or even month. It is important to think of exercise as another important item or task to put on the "to do" list.

Walking is one of the most common forms of physical activity, one that can be done by the vast majority of our society, barring any major physical limitations. Daily walking has previously been linked to important cardiovascular risk variables [117], which were also strongly associated with increased body fat percentage. As mentioned above, individuals can track their daily walking with something as simple as a pedometer. The first electronic pedometer, called "manpo-kei," meaning "10,000 steps meter," was developed by the Japanese" [118]. The well-known walking goal of 10,000 steps per day has been shown to be a reasonable goal of daily activity for healthy adults, and studies have documented the health benefits of attaining similar levels [119, 120]. Those who meet the daily 10,000-step recommendation are more likely to meet physical activity guidelines from the CDC, the ACSM, and the US Surgeon General [121–123]. A recent study found that individuals who accumulate more than 9000 steps per day were more frequently classified as normal weight for height, whereas those accumulating less than 5000 steps per day were more frequently classified as obese [124]. For those who need physical activity the most, the most effective way to eliminate a sedentary lifestyle has been to incorporate pedometers into their activities of daily living [125, 126].

The benefits of using a pedometer revolve around an increase in physical activity. When looking at specific diseases, especially those mentioned in previous sections, there is evidence that they benefit from pedometer use. A meta-analysis of studies that used pedometers to monitor physical activity for weight loss found that using a device increased daily step counts by 2000–4000 steps per day, in addition to weight loss of about 1 lb every 10 weeks [127]. While that might seem modest, over the course of a year, with continued behaviors, it amounts to 5 lbs per year, effectively offsetting the amount of weight gained yearly at now-sedentary careers [3]. Individuals who are looking to lose weight would benefit from incorporating pedometers and activity tracking devices in order to accomplish a goal in a realistic and efficacious manner. Activity tracking has also been useful in people with T2D, as they became more aware of their exercise habits and, as a result, increased their daily activity and exercise capacity [128]. The more exercise minutes a person with T2D accumulates on a regular basis, the more likely they are to eliminate a disease that is often a byproduct of sedentary behavior and obesity. Cancer patients have also found benefit in a pedometer-based walking program, noting a decrease in cancer-related fatigue, improved physical function, and enhanced well-being [129]. Even in the throes of chemotherapy, people with advanced cancer diagnoses felt better performing a little bit more activity than usual for 8 weeks. Results of a study looking at CVD variables and increased activity (tracked by a pedometer) found that there was an increase in steps per day (1500), as well as changes in CVD variables, including HDL, TC, insulin, and a reduction in SBP [130]. The long-term implications of uncontrolled CVD are numerous, and include some very staggering and deleterious results. Given the rather simplistic nature of the necessary exercise to improve CVD values, it makes sense to "prescribe" a way to track activity in order to modify certain risk factors. At this point, it would seem feasible to assume that pedometers can be a useful intervention and motivational device in people with osteoporosis, as sedentary behavior and a lack of weight-bearing activity contribute to the disease. In fact, much has been written about remaining active throughout the lifespan in order to prevent skeletal demineralization from occurring by placing regular stress on one's bones and joints [131]. Therefore, interested parties could extrapolate the results of activity tracking to this population, as well as any other disease or condition that benefits from physical activity that was not specifically focused on to this point. This is not to say there are more or less important afflictions; rather, the ten that were highlighted comprised a large body of research when looking at physical activity and positive outcomes when there was an increase in exercise and/or caloric expenditure. Exercise, in general, promotes a greater sense of physical well-being [132], a by-product even the healthiest person can benefit from.

5. Conclusion

Much of our population lives a sedentary lifestyle. The byproducts of increasing levels of inactivity are quite morose, including fragile bones, poor heart health, and increased body weight. Fortunately, one way to off-set a sedentary lifestyle is to increase daily physical activity. Modest amounts of walking or other forms of cardiovascular exercise (cycling, swimming, hiking) can play a major role in preventing or reversing certain conditions like hypertension, obesity, mortality risk, and T2D. Research has shown that similar conditions can benefit from tracking activity, using pedometers and other commercially available devices that motivate and encourage people to move more and increase their daily levels of activity. Therefore, health-care providers who have the power to recommend such devices and exercise programs should feel empowered to do so. Health and wellness professionals should feel the same sense of empowerment when working with patients and clients who are looking to improve body composition and reduce chronic diseases with increased activity outside of the traditional gym-based workout.

Acknowledgements

The authors would like to thank the Mayo Clinic Healthy Living program for allowing them to diligently work on such an important contribution to academia. Without their support and guidance, they would not be able to put such a thoughtful chapter together. They hope that many will benefit from this information and encourage more people to move.

http://dx.doi.org/10.5772/65240

Author details

Daniel V. Gaz*, Thomas M. Rieck and Nolan W. Peterson

*Address all correspondence to: gaz.daniel@mayo.edu

Mayo Clinic, Rochester, MN, United States

References

[1] Eaton, S.B., An evolutionary perspective on human physical activity: implications for health. Comparative Biochemistry and Physiology. Part A, Molecular & Integrative Physiology, 2003. 136(1): p. 153–9.

[2] Brownson, R.C., T.K. Boehmer, and D.A. Luke, Declining rates of physical activity in the United States: what are the contributors? Annual Review of Public Health, 2005. 26: p. 421–43.

[3] Church, T.S., et al., Trends over 5 decades in U.S. occupation-related physical activity and their associations with obesity. PLoS One, 2011. 6(5): p. e19657.

[4] Matthews, C.E., et al., Amount of time spent in sedentary behaviors in the United States, 2003–2004. American Journal of Epidemiol, 2008. 167(7): p. 875–81.

[5] Exercise: 7 Benefits of Regular Physical Activity. Mayo Clinic, 05 Feb. 2014. Web. 26 Aug. 2016.

[6] Jeffery, R.W., et al., Physical activity and weight loss: does prescribing higher physical activity goals improve outcome? American Journal of Clinical Nutrition, 2003. 78(4): p. 684–9.

[7] Kruk, J., Physical activity in the prevention of the most frequent chronic diseases: an analysis of the recent evidence. Asian Pacific Journal of Cancer Prevention, 2007. 8(3): p. 325–38.

[8] Fox, K.R., The influence of physical activity on mental well-being. Public Health Nutrition, 1999. 2(3A): p. 411–8.

[9] Mock, V., et al., Fatigue and quality of life outcomes of exercise during cancer treatment. Cancer Practice, 2001. 9(3): p. 119–27.

[10] King, A.C., et al., Moderate-intensity exercise and self-rated quality of sleep in older adults. A randomized controlled trial. JAMA, 1997. 277(1): p. 32–7.

[11] Haskell, W.L., et al., Physical activity and public health: updated recommendation for adults from the American College of Sports Medicine and the American Heart Association. Circulation, 2007. 116(9): p. 1081–93.

[12] Garber, C.E., et al., American College of Sports Medicine position stand. Quantity and quality of exercise for developing and maintaining cardiorespiratory, musculoskeletal, and neuromotor fitness in apparently healthy adults: guidance for prescribing exercise. Medicine and Science in Sports and Exercise, 2011. 43(7): p. 1334–59.

[13] One in five adults meet overall physical activity guidelines, in CDC Newsroom. 2013, Centers for Disease Control and Prevention, Atlanta, GA.

[14] Lee, D.C., et al., Mortality trends in the general population: the importance of cardiorespiratory fitness. Journal of Psychopharmacology, 2010. 24(4 Suppl): p. 27–35.

[15] Lee, I.M., et al., Effect of physical inactivity on major non-communicable diseases worldwide: an analysis of burden of disease and life expectancy. Lancet, 2012. 380(9838): p. 219–29.

[16] Durstine, J.L., et al., Chronic disease and the link to physical activity. Journal of Sport and Health Science, 2013. 2(1): p. 3–11.

[17] Sattelmair, J., et al., Dose response between physical activity and risk of coronary heart disease: a meta-analysis. Circulation, 2011. 124(7): p. 789–95.

[18] Sesso, H.D., et al., Physical activity and cardiovascular disease risk in middle-aged and older women. American Journal of Epidemiol, 1999. 150(4): p. 408–16.

[19] Sesso, H.D., R.S. Paffenbarger, Jr., and I.M. Lee, Physical activity and coronary heart disease in men: the Harvard Alumni Health Study. Circulation, 2000. 102(9): p. 975–80.

[20] Rosen, P. "Can You Be Fat But Fit?" Fitness Magazine 1 June 2013: n. pag. Web. 26 Aug. 2016.

[21] MacManus, R. Trackers: How Technology Is Helping Us Monitor & Improve Our Health. Auckland: Bateman, 2014. 55. Print.

[22] Lamkin, P. Fitness tracker market to top $5bn by 2019. Wareable, 2015.

[23] Chiauzzi, E., C. Rodarte, and P. DasMahapatra, Patient-centered activity monitoring in the self-management of chronic health conditions. BMC Medicine, 2015. 13: p. 77.

[24] Bravata, D.M., et al., Using pedometers to increase physical activity and improve health: a systematic review. JAMA, 2007. 298(19): p. 2296–2304.

[25] Pourzanjani, A., T. Quisel, and L. Foschini, Adherent use of digital health trackers is associated with weight loss. PLoS One, 2016. 11(4): p. e0152504.

[26] Cox, J.L., The challenge with tracking health outcomes. Canadian Journal of Clinical Pharmacology, 2001. 8 (Suppl A): p. 10A–16A.

[27] Dobkin, B.H. and A. Dorsch, The promise of mHealth: daily activity monitoring and outcome assessments by wearable sensors. Neurorehabilitation & Neural Repair, 2011. 25(9): p. 788–98.

[28] Hoy, M.B., Personal activity trackers and the quantified self. Medical Reference Services Quarterly, 2016. 35(1): p. 94–100.

[29] Shaw, J.H. and M.R. Rogers, The status of required physical education in colleges and universities of the United States. Research Quarterly, 1946. 17: p. 2–9.

[30] Saunders, R.J., et al., Physical fitness of high school students and participation in physical education classes. Research Quarterly, 1969. 40(3): p. 552–60.

[31] Sallis, J.F., et al., Physical education's role in public health: steps forward and backward over 20 years and HOPE for the future. Research Quarterly for Exercise and Sport, 2012. 83(2): p. 125–35.

[32] Sallis, J.F. and T.L. McKenzie, Physical education's role in public health. Research Quarterly for Exercise and Sport, 1991. 62(2): p. 124–37.

[33] Robb, B. "Exercise and Physical Activity: What's the Difference?" Everyday Health.com. Everyday Health, n.d. Web. 26 Aug. 2016.

[34] "Physical Activity Basics." Centers for Disease Control and Prevention. Centers for Disease Control and Prevention, 04 June 2015. Web. 26 Aug. 2016.

[35] American College of Sports Medicine position stand . The recommended quantity and quality of exercise for developing and maintaining cardiorespiratory and muscular fitness in healthy adults. Medicine and Science in Sports and Exercise, 1990. 22(2): p. 265–74.

[36] The recommended quantity and quality of exercise for developing and maintaining cardiorespiratory and muscular fitness in healthy adults. Position stand of the American College of Sports Medicine . Schweizerische Zeitschrift fur Sportmedizin, 1993. 41(3): p. 127–37.

[37] American College of Sports Medicine Position Stand and American Heart Association . Recommendations for cardiovascular screening, staffing, and emergency policies at health/fitness facilities. Medicine and Science in Sports and Exercise, 1998. 30(6): p. 1009–18.

[38] American College of Sports Medicine Position Stand . The recommended quantity and quality of exercise for developing and maintaining cardiorespiratory and muscular fitness, and flexibility in healthy adults. Medicine and Science in Sports and Exercise, 1998. 30(6): p. 975–91.

[39] Garber, C.E., et al., American College of Sports Medicine position stand. Quantity and quality of exercise for developing and maintaining cardiorespiratory, musculoskeletal, and neuromotor fitness in apparently healthy adults: guidance for prescribing exercise. Medicine and Science in Sports and Exercise, 2011. 43(7): p. 1334–59.

[40] Blair, S.N., et al., Changes in physical fitness and all-cause mortality. A prospective study of healthy and unhealthy men. JAMA, 1995. 273(14): p. 1093–8.

[41] Lee, D.C., et al., Comparisons of leisure-time physical activity and cardiorespiratory fitness as predictors of all-cause mortality in men and women. British Journal of Sports Medicine, 2011. 45(6): p. 504–10.

[42] Sui, X., M.J. LaMonte, and S.N. Blair, Cardiorespiratory fitness as a predictor of nonfatal cardiovascular events in asymptomatic women and men. American Journal of Epidemiology, 2007. 165(12): p. 1413–23.

[43] Sui, X., et al., Cardiorespiratory fitness and adiposity as mortality predictors in older adults. JAMA, 2007. 298(21): p. 2507–16.

[44] Physical Activity Guidelines Advisory Committee report, 2008. To the Secretary of Health and Human Services. Part A: executive summary. Nutrition Reviews, 2009. 67(2): p. 114–20.

[45] Epstein, L.H., K.J. Coleman, and M.D. Myers, Exercise in treating obesity in children and adolescents. Medicine and Science in Sports and Exercise, 1996. 28(4): p. 428–35.

[46] Faigenbaum, A.D., et al., Effects of different resistance training protocols on upper-body strength and endurance development in children. Journal of Strength and Conditioning Research/National Strength & Conditioning Association, 2001. 15(4): p. 459–65.

[47] Vicente-Rodriguez, G., How does exercise affect bone development during growth? Sports Medicine, 2006. 36(7): p. 561–9.

[48] Type 2 diabetes in children and adolescents. American Diabetes Association . Pediatrics, 2000. 105(3 Pt 1): p. 671–80.

[49] Strauss, R.S., et al., Psychosocial correlates of physical activity in healthy children. Archives of Pediatrics & Adolescent Medicine, 2001. 155(8): p. 897–902.

[50] Galland, B.C. and E.A. Mitchell, Helping children sleep. Archives of Disease in Childhood, 2010. 95(10): p. 850–3.

[51] Biddle, S. and N. Armstrong, Children's physical activity: an exploratory study of psychological correlates. Social Science & Medicine, 1992. 34(3): p. 325–31.

[52] "How Much Physical Activity Do Children Need?" Centers for Disease Control and Prevention. Centers for Disease Control and Prevention, 04 June 2015. Web. 26 Aug. 2016.

[53] Physical activity and young people. Global Strategy on Diet, Physical Activity and Health, 2011.

[54] Lees, F.D., et al., Barriers to exercise behavior among older adults: a focus-group study. Journal of Aging and Physical Activity, 2005. 13(1): p. 23–33.

[55] Schutzer, K.A. and B.S. Graves, Barriers and motivations to exercise in older adults. Preventive Medicine, 2004. 39(5): p. 1056–61.

http://dx.doi.org/10.5772/65240

[56] Nelson, M.E., et al., Physical activity and public health in older adults: recommendation from the American College of Sports Medicine and the American Heart Association. Circulation, 2007. 116(9): p. 1094–105.

[57] Shumway-Cook, A., et al., The effect of multidimensional exercises on balance, mobility, and fall risk in community-dwelling older adults. Physical Therapy, 1997. 77(1): p. 46–57.

[58] Falls and Older Age. [cited 2016 July 20]; Available from: http://nihseniorhealth.gov/falls/aboutfalls/01.html.

[59] Guideline for the prevention of falls in older persons. American Geriatrics Society, British Geriatrics Society, and American Academy of Orthopaedic Surgeons Panel on Falls Prevention . Journal of the American Geriatrics Society, 2001. 49(5): p. 664–72.

[60] Kesaniemi, Y.K., et al., Dose-response issues concerning physical activity and health: an evidence-based symposium. Medicine and Science in Sports and Exercise, 2001. 33(6 Suppl): p. S351–8.

[61] Keysor, J.J., Does late-life physical activity or exercise prevent or minimize disablement? A critical review of the scientific evidence. American Journal of Preventive Medicine, 2003. 25(3): p. 129–136.

[62] Pahor, M., et al., Effects of a physical activity intervention on measures of physical performance: results of the lifestyle interventions and independence for Elders Pilot (LIFE-P) study. The Journals of Gerontology. Series A, Biological Sciences and Medical Sciences, 2006. 61(11): p. 1157–65.

[63] Nelson, M.E., et al., The effects of multidimensional home-based exercise on functional performance in elderly people. The Journals of Gerontology. Series A, Biological Sciences and Medical Sciences, 2004. 59(2): p. 154–60.

[64] Oja, P., Dose response between total volume of physical activity and health and fitness. Medicine and Science in Sports and Exercise, 2001. 33(6 Suppl): p. S428–37; discussion S452–3.

[65] Shephard, R.J., Absolute versus relative intensity of physical activity in a dose-response context. Medicine and Science in Sports and Exercise, 2001. 33(6 Suppl): p. S400–18; discussion S419–20.

[66] Warburton, D.E., C.W. Nicol, and S.S. Bredin, Health benefits of physical activity: the evidence. CMAJ : Canadian Medical Association journal = journal de l'Association medicale canadienne, 2006. 174(6): p. 801–9.

[67] Prakash, R. "Physical Inactivity a Leading Cause of Disease and Disability, Warns WHO." WHO. World Health Organization, 04 Apr. 2002. Web. 26 Aug. 2016.

[68] "What Is Metabolic Syndrome?" Health Information for the Public. National Institutes of Health, 26 June 2016. Web. 26 Aug. 2016.

[69] Arem, H., et al., Leisure time physical activity and mortality: a detailed pooled analysis of the dose-response relationship. JAMA Internal Medicine, 2015. 175(6): p. 959–67.

[70] Outdoor Recreation Participation Topline Report, 2016, Outdoor Foundation. p. 10.

[71] Schnohr, P., et al., Dose of jogging and long-term mortality: the Copenhagen City Heart Study. Journal of the American College of Cardiology, 2015. 65(5): p. 411–9.

[72] Mozaffarian, D., et al., Heart disease and stroke statistics-2016 update: a report from the American Heart Association. Circulation, 2016. 133(4): p. e38–60.

[73] Miller, SG. "How Much Do Chronic Diseases Cost in the US?" Live Science. TechMedia Network, 03 Sept. 2015. Web. 26 Aug. 2016.

[74] Buttar, H.S., T. Li, and N. Ravi, Prevention of cardiovascular diseases: role of exercise, dietary interventions, obesity and smoking cessation. Experimental and Clinical Cardiology, 2005. 10(4): p. 229–49.

[75] Thompson, P.D., et al., Exercise and physical activity in the prevention and treatment of atherosclerotic cardiovascular disease: a statement from the Council on Clinical Cardiology (Subcommittee on Exercise, Rehabilitation, and Prevention) and the Council on Nutrition, Physical Activity, and Metabolism (Subcommittee on Physical Activity). Circulation, 2003. 107(24): p. 3109–16.

[76] Global Report on Diabetes, 2016, World Health Organizatoin: Geneva.

[77] Mathers, C.D. and D. Loncar, Projections of global mortality and burden of disease from 2002 to 2030. PLoS Medicine, 2006. 3(11): p. e442.

[78] Zinman, B., et al., Physical activity/exercise and diabetes mellitus. Diabetes Care, 2003. 26 Suppl 1: p. S73–7.

[79] Tuomilehto, J., et al., Prevention of type 2 diabetes mellitus by changes in lifestyle among subjects with impaired glucose tolerance. New England Journal of Medicine, 2001. 344(18): p. 1343–1350.

[80] Pan, X.R., et al., Effects of diet and exercise in preventing NIDDM in people with impaired glucose tolerance. The Da Qing IGT and Diabetes Study. Diabetes Care, 1997. 20(4): p. 537–44.

[81] Group, D.P.P.R., Reduction in the incidence of type 2 diabetes with lifestyle intervention or metformin. The New England Journal of Medicine, 2002. 2002(346): p. 393–403.

[82] Eriksson, K.F. and F. Lindgarde, Prevention of type 2 (non-insulin-dependent) diabetes mellitus by diet and physical exercise. The 6-year Malmo feasibility study. Diabetologia, 1991. 34(12): p. 891–8.

[83] Knowler, W.C., et al., Reduction in the incidence of type 2 diabetes with lifestyle intervention or metformin. The New England Journal of Medicine, 2002. 346(6): p. 393–403.

[84] Ladabaum, U., et al., Obesity, abdominal obesity, physical activity, and caloric intake in US adults: 1988 to 2010. The American Journal of Medicine, 2014. 127(8): p. 717–727 e12.

[85] Serdula, M.K., et al., Do obese children become obese adults? A review of the literature. Preventive Medicine, 1993. 22(2): p. 167–177.

[86] Ross, R. and I. Janssen, Physical activity, total and regional obesity: dose-response considerations. Medicine and Science in Sports and Exercise, 2001. 33(6 Suppl): p. S521–7; discussion S528–9.

[87] Mahabadi, A.A., et al., Association of pericardial fat, intrathoracic fat, and visceral abdominal fat with cardiovascular disease burden: the Framingham Heart Study. European Heart Journal, 2009. 30(7): p. 850–6.

[88] Sahakyan, K.R., et al., Normal-weight central obesity: implications for total and cardiovascular mortality. Annals of Internal Medicine, 2015. 163(11): p. 827–35.

[89] Lee, I.M., Physical activity and cancer prevention—data from epidemiologic studies. Medicine and Science in Sports and Exercise, 2003. 35(11): p. 1823–7.

[90] Meyerhardt, J.A., et al., Physical activity and survival after colorectal cancer diagnosis. Journal of Clinical Oncology: Official Journal of the American Society of Clinical Oncology, 2006. 24(22): p. 3527–34.

[91] Sallis, R.E., Exercise is medicine and physicians need to prescribe it! British Journal of Sports Medicine, 2009. 43(1): p. 3–4.

[92] Baster, T. and C. Baster-Brooks, Exercise and hypertension. Australian Family Physician, 2005. 34(6): p. 419–24.

[93] "High Blood Pressure Dangers: Hypertension's Effects on Your Body." Diseases and Conditions. Mayo Clinic, 18 Feb. 2014. Web. 26 Aug. 2016.

[94] 10 ways to control high blood pressure without medication. Patient Care & Health Info, 2015.

[95] Kokkinos, P.F. and V. Papademetriou, Exercise and hypertension. Coronary Artery Disease, 2000. 11(2): p. 99–102.

[96] Ray, N.F., et al., Medical expenditures for the treatment of osteoporotic fractures in the United States in 1995: report from the National Osteoporosis Foundation. Journal of Bone and Mineral Research: The Official Journal of the American Society for Bone and Mineral Research, 1997. 12(1): p. 24–35.

[97] Phy, M.P., et al., Effects of a hospitalist model on elderly patients with hip fracture. Archives of Internal Medicine, 2005. 165(7): p. 796–801.

[98] O'Brien, M., Exercise and osteoporosis. Irish Journal of Medical Science, 2001. 170(1): p. 58–62.

[99] Charansonney, O.L., Physical activity and aging: a life-long story. Discovery Medicine, 2011. 12(64): p. 177–85.

[100] Sinaki, M., Exercise and osteoporosis. Archives of Physical Medicine and Rehabilitation, 1989. 70(3): p. 220–9.

[101] Berg, A., et al., Physical activity and lipoprotein lipid disorders. Sports Medicine, 1994. 17(1): p. 6–21.

[102] Durstine, J.L. and P.D. Thompson, Exercise in the treatment of lipid disorders. Cardiology Clinics, 2001. 19(3): p. 471–88.

[103] Durstine, J.L., et al., Lipids, lipoproteins, and exercise. Journal of Cardiopulmonary Rehabilitation, 2002. 22(6): p. 385–98.

[104] Depression (major depressive disorder). Patient Care & Health Info, 2016.

[105] Nabkasorn, C., et al., Effects of physical exercise on depression, neuroendocrine stress hormones and physiological fitness in adolescent females with depressive symptoms. European Journal of Public Health, 2006. 16(2): p. 179–84.

[106] Brosse, A.L., et al., Exercise and the treatment of clinical depression in adults: recent findings and future directions. Sports Medicine, 2002. 32(12): p. 741–60.

[107] North, T.C., P. McCullagh, and Z.V. Tran, Effect of exercise on depression. Exercise and Sport Sciences Reviews, 1990. 18: p. 379–415.

[108] Lawlor, D.A. and S.W. Hopker, The effectiveness of exercise as an intervention in the management of depression: systematic review and meta-regression analysis of randomised controlled trials. BMJ, 2001. 322(7289): p. 763.

[109] Generalized anxiety disorder. Patient Care & Health Info: Diseases and Conditions, 2014.

[110] Petruzzello, S.J., et al., A meta-analysis on the anxiety-reducing effects of acute and chronic exercise. Sports Medicine, 1991. 11(3): p. 143–182.

[111] Facts & Statistics: Did You Know. About ADAA 2014 September 2014 [cited 2016 July 19, 2016]; Available from: http://www.adaa.org/about-adaa/press-room/facts-statistics.

[112] McAuley, E., S.L. Mihalko, and S.M. Bane, Acute exercise and anxiety reduction: does the environment matter? Journal of Sport and Exercise Psychology, 1996. 18: p. 408–419.

[113] Petruzzello, S.J., and D.M. Landers. "State Anxiety Reduction and Exercise." Medicine and Science in Sports and Exercise 26.8 (1994): 1028–035.

[114] Bahrke, M.S. and W.P. Morgan, Anxiety reduction following exercise and meditation. Cognitive Therapy and Research, 1978. 2(4): p. 323–333.

[115] Thompson, D.L., et al., Pedometer-measured walking and risk factors for disease. Southern Medical Journal, 2006. 99(1): p. 100–1.

[116] Fox, S., Health topics. Pew Research Center's Internet & American Life Project, 2011, Retirado de http://www. pewinternet.org/2011/02/01/health-topics-2.

[117] Oja, P., I. Vuori, and O. Paronen, Daily walking and cycling to work: their utility as health-enhancing physical activity. Patient Education and Counseling, 1998. 33: p. S87–S94.

[118] Hatano, Y., Use of the pedometer for promoting daily walking exercise. International Council for Health, Physical Education, and Recreation, 1993. 29(4): p. 4–8.

[119] Tudor-Locke, C., et al., Controlled outcome evaluation of the First Step Program: a daily physical activity intervention for individuals with type II diabetes. International Journal of Obesity and Related Metabolic Disorders: Journal of the International Association for the Study of Obesity, 2004. 28(1): p. 113–9.

[120] Tudor-Locke, C., et al., Revisiting ``how many steps are enough?''. Medicine and Science in Sports and Exercise, 2008. 40(7): p. S537.

[121] Le Masurier, G.C., C.L. Sidman, and C.B. Corbin, Accumulating 10,000 steps: does this meet current physical activity guidelines? Research Quarterly for Exercise and Sport, 2003. 74(4): p. 389–94.

[122] Pate, R.R., et al., Physical activity and public health. A recommendation from the Centers for Disease Control and Prevention and the American College of Sports Medicine. JAMA, 1995. 273(5): p. 402–7.

[123] Shalala, D.E. "Physical Activity and Health: A Report of the Surgeon General." Centers for Disease Control and Prevention. Centers for Disease Control and Prevention, 17 Nov. 1999. Web. 26 Aug. 2016.

[124] Tudor-Locke, C., et al., The relationship between pedometer-determined ambulatory activity and body composition variables. International Journal of Obesity & Related Metabolic Disorders, 2001. 25(11): 1571–1578.

[125] Hultquist, C.N., C. Albright, and D.L. Thompson, Comparison of walking recommendations in previously inactive women. Medicine and Science in Sports and Exercise, 2005. 37(4): p. 676–83.

[126] Stovitz, S.D., et al., Pedometers and brief physician counseling: increasing physical activity for patients seen at a family practice clinic. Medicine & Science in Sports & Exercise, 2004. 36(5): p. S241.

[127] Richardson, C.R., et al., A meta-analysis of pedometer-based walking interventions and weight loss. Annals of Family Medicine, 2008. 6(1): p. 69–77.

[128] Korkiakangas, E. et al. "Pedometer Use Among Adults at High Risk of Type 2 Diabetes, Finland, 2007–2008." Preventing Chronic Disease 7.2 (2010): A37. Print.

[129] Mayo, N.E., et al., Pedometer-facilitated walking intervention shows promising effectiveness for reducing cancer fatigue: a pilot randomized trial. Clinical rehabilitation, 2014. 28(12): p. 1198–209.

[130] Miyazaki, R., et al., Effects of a year-long pedometer-based walking program on cardiovascular disease risk factors in active older people. Asia-Pacific Journal of public health/Asia-Pacific Academic Consortium for Public Health, 2015. 27(2): p. 155–63.

[131] Drinkwater, B.L., Exercise in the prevention of osteoporosis. Osteoporosis International : A Journal Established as Result of Cooperation between the European Foundation for Osteoporosis and the National Osteoporosis Foundation of the USA, 1993. 3 Suppl 1: p. 169–71.

[132] Hassmen, P., N. Koivula, and A. Uutela, Physical exercise and psychological well-being: a population study in Finland. Preventive Medicine, 2000. 30(1): p. 17–25.

Chapter 4

Neurophysiological Effects of Exercise

Selma Cırrık and Gulay Hacioglu

Additional information is available at the end of the chapter

http://dx.doi.org/10.5772/64801

Abstract

Convincing findings from animal and clinical studies have shown that exercise improves mood and cognition in addition to cardiovascular and metabolic benefits. Exercise, with the greatest effects on the hippocampus, which has a central role in learning and memory, increases neurogenesis and synaptic plasticity. Although the exact molecular mechanisms responsible for the exercise-induced neuroplasticity need to be clarified, some neurotrophic and angiogenic factors (e.g. BDNF, IGF-1, bFGF2 and VEGF) and different neurotransmitter systems (glutamate, GABA, endocannabinoids and monoamines) may have critical contributions in these processes. Exercise-induced changes in the brain morphology, chemistry and functions seem to be responsible for the beneficial effects of exercise, like improved learning and memory, anti-depressant-like and anxiolytic effects, reduced cognitive decline related to ageing and improvements in symptoms of neurodegenerative diseases. In this chapter, after discussing basic neurophysiological information regarding the brain, cognition, neurotransmitter systems, neural plasticity, learning, memory and behaviour tasks, the focus is on the exercise-induced changes in neuroplasticity, cognitive functions and mood and the factors mediating the effects of exercise, and finally, the effect of exercise on ageing and neurodegenerative diseases is discussed.

Keywords: exercise, cognition, learning, memory, depression, ageing, neurodegenerative diseases

1. Introduction

The brain is highly dynamic, constantly reorganizing organ that undergoes both acute and chronic changes throughout the lifespan. There is a remarkable linkage between structural and functional brain modulations and exercise, as several recent papers have comprehensively described the neuromolecular events resulting from regular exercise (will be further discussed in the text). Though physical activity and exercise are commonly used in place of the other, there are two terms with different meanings. Physical activity is defined as body movements by skeletal muscles resulting in energy consumption, while exercise is defined as a planned, structured and repeated type of physical activity. The cardiovascular and metabolic benefits of regular exercise are well known, and it causes a reduction in risk for many diseases such as obesity, type 2 diabetes, heart diseases, cancer and other diseases and chronic conditions [1, 2]. Studies have shown that exercise is not just effective on peripheral tissue, at the same time, it affects the central nervous system (CNS). for example, regular exercise improves cognition, reduces age-related memory loss, delays and alleviates the symptoms of neurodegenerative diseases, increases the speed of neuronal healing after injury and is even known to improve depression. However, the underlying mechanisms responsible for the beneficial effect of exercise are still unclear, a variety of cellular and molecular systems important for maintaining neuronal network and synaptic function such as neurotransmitters, angiogenic and neurotrophic factors may be instrumental for positive effects of exercise on brain [3–7]. Nevertheless, growing evidence suggests that an increase in hippocampal neurogenesis may mediate, at least in part, the exercise-induced improvement in cognitive function and mood.

2. Basic neurophysiology

The brain is one of the most important and magnificent organs of an organism that can be categorized under the CNS. The human brain contains nearly 98% of the body's neural tissue [8]. It is made up of more than 100 billion nerves that correspond with one another by means of long protoplasmic fibres called axons, which carry signal pulses—action potentials—to distant parts of the brain or body targeting specific recipient cells [9]. Physiologically, the function of the brain is to exercise centralized control over the other organs of the body. It controls and coordinates most sensory systems, muscle movements, social behaviour and homoeostatic body functions such as heart and respiratory rate, blood pressure, fluid balance and body temperature. The brain is the source of cognition, mood, emotion, memory, motor and other forms of learning.

2.1. The structure and functional organization of the brain

The adult brain is dominated in size by the cerebrum. The cerebrum is the seat of most higher mental functions such as conscious thoughts, sensations, intelligence, memory and complex movements. It can be divided into two main parts: the right and left cerebral hemispheres. The surfaces of the cerebral hemispheres are highly folded and covered by neural cortex, a superficial layer of grey matter, with an inner layer of white matter [8]. Grey matter consists mainly of neuronal cell bodies, dendrites, glial cells (astroglia and oligodendrocytes), synapses and capillaries, while white matter consists primarily of axons that connect various areas of the brain [10, 11]. Each hemisphere provides a different set of activities, behaviours and controls. The right hemisphere is often called the creative side of the brain, while the left hemisphere is the rational or analytical side of the brain. The two hemispheres communicate through a large bundle of nerve fibres referred as *corpus callosum* and through several smaller nerve pathways [10]. Therefore, even though there are some functions that seem to be dominant in one hemisphere, both hemispheres play a part in overall brain activity.

The second largest part of the brain is cerebellum, which is partially hidden by the cerebral hemispheres [8]. As the cerebral cortex, the cerebellum consists of white matter and a thin, outer layer of densely folded grey matter. The cerebellum processes information from the brain and peripheral nervous system. It is involved in several functions such as fine movement coordination, balance and equilibrium, muscle tone and sense of body position.

Generally, there are four recognized sections in the brain, known as lobes (frontal, occipital, parietal and temporal). While any complex ability depends on the coordinated process of neural networks across lobes, each lobe can be roughly linked with particular functions [10]. The frontal lobe lies beneath the forehead, and it is responsible for cognitive and reasoning functions, social behaviour and motor skills. The parietal lobe is posterior to the frontal lobe and has several functions including sensation, perception and spatial reasoning. This lobe is responsible for interpreting sensory information from various parts of the body. The smallest lobe of the four lobes, the occipital lobe, is located at the back of the brain, and it is the primary visual–spatial processing centre. Lastly, the temporal lobe is located on both sides of the brain, which is involved in the processing of auditory input (sound) and the olfactory sense (smell). Each lobe is further subdivided into interlinking networks of neurons specialized for very precise information processing. These four regions of the brain work in harmony in order to maintain the homoeostatic balance of the human body.

2.2. Brain development

Brain development results from the common work of genes and environment, where genes have the function of leading the initial steps of development and the generation of neural connections and networks [12]. The very first step of brain development stage begins with the recruitment of progenitor cells (precursors of neurons and glial cells) into a specialized structure called the neural plate. The neural plate then folds to form the neural tube, an early embryonic structure. As the foetus develops, the grooves and plicas in the neural tube intensify, producing particular layers of the brain. From the anterior part of the neural tube generates the telencephalon, which enlarges rapidly due to cell division and proliferation, and eventually gives rise to the brain. Some of the cells differentiate into neuronal and glial cells, which are the major cellular constituents of the brain. The newly produced neurons migrate to different areas of the developing brain to self-organize into main brain structures. When the neurons reach their regional positions, they extend axons and dendrites, which provide them communication with other neurons via synapses. Synaptic intercommunication leads to the establishment of specific neural lines that mediate sensory and motor processing and underlie behaviour. The first areas of the brain to fully develop are the brainstem and midbrain, responsible for the autonomic functions necessary for life. At birth, these lower segments of the nervous system are effectively developed, whereas the higher regions like the limbic system and the cerebral cortex are still rather primitive. Brain regions involved in complicated neuronal functions such regulating emotions, cognition and language grow rapidly in the first

3 years, and then, the brain proceeds its development gradually within the first 20 years of life [10, 11].

Studies using high-resolution imaging techniques demonstrate that the brain still continues to grow and develop undergoing notable changes into adulthood, as white matter volume has been shown to increase in adults [13, 14]. The process of adult neurodevelopment includes several mechanisms such as formation of new brain cells, synapses and cellular components, changes in cerebral blood flow and neuronal activity, as well as myelination of axons. Neurodevelopment takes place both in the central and peripheral nervous system, although there are some differences in terms of speed and extent.

2.3. Introduction to main neurotransmitters

Neurotransmitters are endogenous chemicals, which modulate signals and provide efficient communication among large numbers of brain cells. They are synthesized by neurons and packaged into synaptic vesicles that cluster beneath the presynaptic membrane of a synapse and are released into the synaptic cleft, where they bind to specialized receptors in the postsynaptic membrane of the synapse, changing its activity [15]. The major neurotransmitters of the brain are summarized below.

2.3.1. Glutamate

Glutamate is the most common neurotransmitter in the brain. It always has an excitatory role. Most glutamate is produced in neurons from glutamine, which is mainly formed in astrocytes and transported into neurons [16]. There are several types of glutamate receptors:

2.3.1.1. AMPA receptors

These are ionotropic receptors (ligand-gated ion channels). Their opening allows the influx of calcium and sodium and the efflux of potassium.

2.3.1.2. N-methyl-D-aspartate (NMDA) receptors

These are also ionotropic receptors increasing the permeability for calcium, sodium and potassium. The opening of NMDA receptors needs a conformational change for the removal of a magnesium ion, which closes the pore. If the membrane is sufficiently depolarized via AMPA receptors, magnesium leaves the channel. NMDA receptors allow a calcium influx adequate for the function of calcium-dependent enzymes, which can then regulate properties of the synapse, thereby forming the synaptic plasticity. The receptors are densely centralized in the hippocampus, basal ganglia and amygdala.

2.3.1.3. Kainate receptors

Kainate receptors are also ionotropic and functionally similar to AMPA receptors.

http://dx.doi.org/10.5772/64801

2.3.1.4. Metabotropic receptors

Eight well-known types of metabotropic receptors are identified as mGluR1-8. They are mostly found on presynaptic neurons, where they increase the turnover of phosphatidylinositol.

2.3.2. GABA (γ-aminobutyric acid)

Occurring in 30–40% of all synapses of the brain GABA is the main inhibitory neurotransmitter. It is synthesized from glutamate by glutamic acid decarboxylase [15]. GABAergic receptors are as follows:

2.3.2.1. GABA-A

An ionotropic receptor allows the influx of chloride into the cytoplasm, thereby leading to a lowering of the membrane potential. The receptor has a supramolecular complex design, which in addition to GABA binds benzodiazepines, barbiturates, corticosteroids and alcohol.

2.3.2.2. GABA-B

This is a metabotropic receptor inhibiting adenylate cyclase via $G\text{-}_{\text{inhibitory}}$ protein. Thus, decrease in cAMP levels mediates membrane protein phosphorylation status, which induces a hyperpolarization and a decreased activity of calcium channels.

2.3.2.3. GABA-C (GABA-A-rho receptor)

An ionotropic receptor activation leads to an influx of chloride ions into the cytoplasm. In contrast to GABA-A receptors, they open more slowly and remain open for a longer period of time.

2.3.3. Dopamine

Dopamine in the CNS plays an important role in the regulation of motor functions, initiation of behavioural patterns and modulation of visceral functions. Five dopamine receptors have been identified, which are all associated with a G-protein regulating adenylate cyclase activity. They can be divided into two groups depending on whether they activate adenylate cyclase via a $G\text{-}_{\text{stimulatory}}$ protein (D1, D5) or inhibit it via a $G\text{-}_{\text{inhibitory}}$ protein (D2–D4). The concentration of receptor subtypes varies in different areas of the CNS; for instance, cortical motor areas are rich in D2, while the limbic system mainly expresses D3 and D4 receptors. Dopaminergic neurons are widely distributed in the hypothalamus and mesencephalon [15].

2.3.4. Noradrenaline

Noradrenaline in the CNS mainly regulates the activity of other neurotransmitters. Generally, noradrenergic pathways modulate both the excitatory function of glutamate and the inhibitory function of GABA. Noradrenergic neurons are found in the brainstem, particularly in the locus coeruleus, tegmentum and the reticular formation of the medulla and pons [15].

2.3.5. *Serotonin*

Serotonin is an indolamine which plays a similar role in the CNS as the noradrenaline. It regulates a range of activities and mediates the function of other projection circuits. Seven models of serotonin receptors have been identified to date, 5-HT1-7R, some of which are excitatory and some inhibitory. The effect of serotonin depends on their localization and the types of receptors presented [15]. Serotonergic neurons can be mainly found in raphe nuclei of the reticular formation.

2.3.6. *Acetylcholine (ACh)*

ACh is a neurotransmitter used in neuromuscular junctions of all vertebrates, all preganglionic neurons of the autonomous nervous system and all postganglionic parasympathetic neurons. In the CNS, it regulates several cortical activities such as arousal, sleep and memory consolidation. ACh is synthesized by transferring the acetyl group from acetyl-CoA to choline by choline acetyltransferase. The cholinergic signal is terminated by the serine hydrolase acetylcholine esterase, bound to the postsynaptic membrane. The hydrolysis produces choline, which is taken up by the presynaptic neuron and recycled, and acetate [15].

2.4. Cognition

Cognition is the number of abilities that is related to knowledge, understanding, thinking and awareness. The Latin root of cognition is *cognoscene*, which translates into "to conceptualize", "to recognize" and "to know". Cognitive abilities cover processes such as attention, memory, judgement and evaluation, logic and reasoning, problem solving and decision-making, comprehension, production of language and the application of acquired knowledge. It is important to realize that these processes are overlapping in nature and often work together in complex ways to formulate any conclusions about the external and internal world. In healthy individuals, the brain is capable of learning new skills in each of these areas [17].

The cognitive ability of humans is the most advanced in the entire animal kingdom. Such advancement is believed to be conferred by an expanded cerebral cortex and a highly developed prefrontal cortex, both of which are brain regions important for cognition. Cognition changes throughout the lifespan matching the development, maturation and ageing of the brain. While cognitive activity gains its peak during middle adulthood, there is a normal degradation in cognitive ability and brain atrophy with age [18, 19]. The revelation that physical changes in the brain can occur across the lifespan provides a biological basis supporting the importance of neuromodulators, cognitive trainings and the efficacy of physical activity [20].

2.5. Neural plasticity

Neural plasticity can be generally defined as the ability of the nervous system to embrace a new functional or structural condition in response to various extrinsic and intrinsic determinants [21]. The term "neuronal plasticity" date back to Santiago Ramon y Cajal (1852–1934) who described non-pathological changes in the structure of adult brains [22]. In a wider sense, plasticity of the brain can be regarded as "the ability to make adaptive changes related to the structure and role of the nervous system at the cellular, molecular, and system levels" [22, 23]. Neuronal plasticity can refer not only for morphological changes in certain brain areas, for alterations in neuronal networks including changes in neuronal connectivity, as well as the formation of new neurons (*neurogenesis*), but also for neurobiochemical changes (*synaptic or non-synaptic plasticity*) [22]. Neuroscientists distinguish synaptic plasticity, which points to alterations in how neurons link up to each other, from non-synaptic plasticity, which refers to changes in the neurons themselves. Neuroplastic modulation can occur at small ranges, such as physical changes to individual neurons, or at whole-brain scales, such as cortical remapping in feedback to injury.

Neuronal plasticity in the brain is greatly enhanced during critical periods early in life and was long believed to be limited thereafter. Studies from a number of laboratories investigating the reorganizations in somatosensory, primary visual, primary auditory cortices and in thalamus have unwrapped a considerable degree of plasticity in the mature brain, too [24]. It is now accepted that the brain has an incredible potential to modify its morphological and functional organization throughout the lifespan, in response to changes in environmental inputs. This brain plasticity underlies normal development and maturation, skill learning and memory, healing from injury, as well as the outcomes of sensory deprivation or environmental enrichment [25]. Usually, plasticity in the adult neocortex lies silent, but can be reactivated by modifications of sensory input or sensory-motor interferences, which alter the level and pattern of activity in cortical circuits [26]. Such treatments, potentially in combination with drugs targeting molecular brakes on plasticity present in the mature brain, might help recovery of activity in the injured or diseased brain [26].

The molecular mechanisms of brain plasticity are under intensive scrutinizations. Calcium ions and channels, glutamate and NMDA receptors, free radicals, lipid peroxides and growth factors (neurotrophins) play a crucial role in these processes. A form of functional neuronal plasticity is long-term potentiation (LTP) that is the long-lasting enhancement in signal transmission between two neurons after synchronous stimulation [22, 27].

2.5.1. Neurogenesis

The most appealing phenomenon of neuroplasticity appears to be adult neurogenesis. Adult neurogenesis occurs in the certain brain areas. So far it appears that in most mammals, the formation of new neurons in adult brains occurs in two parts, the subventricular zone and the dentate gyrus (DG) of hippocampus (hippocampal neurogenesis), and the number of newly generated neurons is fewer compared to the total number of brain cells [22]. However, there are also reports from studies in mice that new neurons can be gradually generated in the adult substantia nigra (SN) [22, 28]. The existence of neurogenesis in mature brains rises hope that even injured brain regions can be functionally repaired and reactivated. Damage to the adult brain such as ischaemic insults causes the proliferation of subventricular zone cells and thus the generation of neuronal precursor cells. These precursors migrate along blood vessels to the wounded region [22, 29]. Nevertheless, only a small percentage can survive, because inflammatory effects that occur in the ischaemic brain area inhibit neurogenesis and the proper integration of new cells into a functional neuronal network [22, 30]. Anti-inflammatory drugs can strengthen neurogenesis, as observed in rodent studies of peripheral inflammation [22, 31].

2.5.2. *Synaptic plasticity*

Synaptic plasticity is an essential organizational characteristic of mammalian brain function, and it refers to the condition where existing connections among neurons are reinforced or weakened and new synapses are formed or existing ones removed. The potential for synaptic plasticity and therefore for learning and memory is not steady throughout life, as it often reaches a peak soon after birth and then gradually declines with increasing age [26]. Besides, synaptic plasticity in the adult brain is widely distributed and is a key feature of various brain regions, like the hippocampus, the striatum or the cerebellum. Mechanisms of synaptic plasticity are a modification of excitation and inhibition process; a change in LTP or long-term depression (LTD); a modulation of neuronal excitability; and the anatomical changes, which require a longer period of time.

2.6. Neural base of learning and memory

Learning and memory are two intimately linked cognitive processes that derive from interactions with the environment. Learning is defined as the act of acquiring new knowledge or skills; it also may involve a change in attitude or behaviour through experience, instruction or study, whereas memory is a cognitive process of the brain enabling past experiences and information to be remembered and recalled. Memory is built on learning and is subject to the same factors influencing learning. This is why memorization of an event or of information can be improved by a strong emotional situation, a special state, high motivation or elevated attention [10]. Learning and memory undergo significant change over the lifespan in both males and females. The ability to learn and remember begins early during the developmental period, is moulded by environmental and hormonal influences during puberty, reaches its apex during adulthood and finally declines with advanced age [32].

The nature of the cellular basis of learning and memory remains obscure despite being one of the most densely examined nervous system mechanisms. A popular model for the physiological processes underlying learning and memory involves that memories are collected by alterations in the strength of synaptic connections within the appropriate neural circuitry. Thus, synaptic plasticity is considered a fundamental working mechanism of memory and learning in biological neural networks. Several lines of evidence have converged to indicate that learning and memory setting requires plasticity of dendritic spines in the medial prefrontal cortex and the hippocampus [32]. Major characteristic of the hippocampus is the ability of hippocampal neurons to undergo LTP, a persistent increase in synaptic intensity emerging from long-lasting electrical stimulation.

2.6.1. *Hippocampus*

The hippocampus is the region in the brain most closely associated with learning. Located inside the medial temporal lobe, beneath the cortical surface of the brain, it belongs to the limbic system and plays important activity in learning, memory and spatial navigation. Thus, damage to the hippocampus can produce profound memory impairments, specifically in the ability to create long-term memories. The names of the main histological divisions of the hippocampus are DG, CA1, CA2 and CA3 regions [33]. The entorhinal cortex, the greatest source of hippocampal input and target of hippocampal output, is powerfully and reciprocally associated with many other parts of the cerebral cortex and therefore acts as the major link between the hippocampus and other parts of the brain. Within the hippocampus, the flow of signals and information is generally unidirectional, first to the DG, then to the CA3 part, then to the CA1 part, then to the subiculum and then out of the hippocampus to the entorhinal cortex [33, 34].

2.6.2. Long-term potentiation

LTP is a physiological model to explain the formation of definite forms of learning and memory, and it refers to changes in a cell that cause it to respond more efficiently to stimulation [35]. It was first described in detail by Bliss and Lomo in 1973, reporting that repetitious activation of excitatory synapses in the hippocampus induced an increment in synaptic strength that generally last for hours or even days [36]. Since then, LTP has been studied intensively over the years, and a great deal has been learned about it. The best-studied form of LTP occurs at synapses that come to an end on dendritic spines and use the transmitter glutamate. The synaptic changes depend on special type of glutamate receptor, the ionotropic NMDA receptor, which possesses the special characteristic of permitting calcium ions to get into the postsynaptic spine only when presynaptic activation and postsynaptic depolarization occur at the same time [37]. LTP experiments have focused especially on the three main pathways in the hippocampal network: the perforant pathway between the entorhinal cortex and DG granule cells, the mossy fibre pathway between DG granule cells and CA3 pyramidal cells, and the Schaffer collateral pathway between CA3 and CA1 pyramidal cells. NMDA receptor is the trigger for the induction of LTP at synapses made between pyramidal neurons in CA3 and CA1 layers of the hippocampus [38]. Drugs that interfere with NMDA receptors prevent LTP formation and also have negative effects on some types of memory, especially spatial memory. Although NMDA receptors display a significant role in synaptic plasticity, thus in learning and memory, sustained NMDA receptor activation can lead to various pathological conditions [39–41]. Therefore, too much activation of the NMDA receptor is detrimental.

2.7. Experimental behaviour tasks

Spatial learning and memory are neurobiological activities that allow us to remember important details related to our surroundings. Researches evaluate this phenomenon in rodents using different types of behavioural tasks. By investigating spatial memory in experimental tasks, neurobehavioral scientists can gain valuable understanding of how these processes are vary after brain injury in humans or how the functions of particular genes on learning are affected. The tests most widely used and for which the most data exist are briefly described here.

2.7.1. T mazes

T mazes are the first labyrinths that can be used to measure spatial working memory. There are many test procedures; however, the easiest to evaluate spatial learning and memory is to prize rodents for rotating in one direction across a series of experiments. The T-maze is shaped like a T. The test animal begins from the base of the T labyrinth. A reward may be placed at one of the arms of the maze or different rewards may be placed at each arm. The rat/mouse walks forward and chooses the left or right arm of the maze. On the test trial, to try out if the animals learn a position habit such as always turning right, the maze is rotated 180°, and if it is a habitual behaviour, the animal will turn right regardless of where that arm is within the room, but if it is a learned behaviour, the rodent will turn left so as to end up in the same place in relation to distal cues [42].

2.7.2. Morris water maze

The Morris water maze is consisting of a round tank filled with opaque water with two small hidden platforms located 1–2 cm under the surface of the water [43]. As soon as the rodent is placed on a start platform, it begins to swim around until it finds the other platform to stand on. The platform is not visible, so the animals must depend on their spatial memory and use extra-maze visual cues to locate the platform. The animal is usually introduced into the water at different locations around the tank. The researcher measures how long it takes for a rodent to find hidden platform. Data collection in the water maze can be simply optimized and automated, and it does not need food-restriction. In addition to this, the water maze task does require complex and coordinated movements [44].

2.7.3. Radial arm maze

Among behavioural tests, one of the most suitable devices for measuring spatial learning and memory is the radial arm maze [45, 46]. Briefly, radial arm maze consists of eight horizontal arms placed radially around a central platform above the floor. Typically, a cup at the end of several of the arms contains some small nutrient reward for the experimental subject. The rodents are usually food deprived to motivate the foraging behaviour, and they will try to find the food in the most effective way. The optimal strategy is to visit each arm once before returning to any formerly visited branch. Therefore, to complete the experiment successfully, animals must remember where it has been and where food has been and/or remains available [47].

3. Exercise-induced changes in the brain

3.1. Exercise and neuroplasticity

Neuroplasticity or brain plasticity refers to modification of the nervous system in response to differing needs or environmental conditions. Currently, scientific information indicates that neuroplasticity extends beyond the developmental period into the adult period and plays a significant role in processes such as learning and gaining new skills, consolidation and recall of memories as well as healing after injury. Neuroplasticity is mediated by different mechanisms causing both functional and morphological changes in the CNS-like neurogenesis, apoptosis, increase and/or decrease in synaptic activity and reorganization of the neuronal network [3–6]. Of these, though neurogenesis is a mechanism only accepted since the 1960s, currently it is well known that newborn neurons can be formed in the olfactory bulb and DG of the adult mammalian brains [3–6].

As mentioned before, the hippocampus has a unique role in learning and memory formation, with high capacity for neuroplasticity. There are many studies showing exercise strongly stimulates neuroplasticity in the hippocampus [1–5]. Rearrangement of DG morphology, especially, in response to exercise is noteworthy, with increases in total length and complexity of granular cells, spine density in dendrites and neurogenesis. These effects of exercise are not limited to the DG, but include entorhinal cortex and CA1 pyramidal cells [3, 5, 48]. However, it should be noted that exercise-induced neurogenesis occurs specifically in DG subfield of hippocampus [4–6]. Borght et al. showed that 1–10 days of wheel-running exercise induces a gradually increment in the hippocampal cell proliferation [49]. Although proliferation had returned to baseline level 1 day after cessation of running, the number of immature neurons remained in the high level for at least 6 days. In rats, a single bout of resistance exercise increases synapsin I, synaptophysin and PSD-95 (a protein related to synapses) levels in the hippocampus and improves contextual memory [50]. In addition to the increase in neuron numbers with exercise, importantly, enhanced maturation of newborn neurons has been also reported [48]. Therefore, these findings suggest that the hippocampus, especially DG, has a considerable capacity for neuroplasticity and rapidly responds to alterations in physical activity.

The increased hippocampal plasticity with exercise is accompanied by an enhancement in synaptic activity of neurons. It is known that the physiologic model of learning and memory of LTP is affected by exercise. Running induces an increase in LTP amplitude in the DG region, but did not cause a change in CA1 region of the hippocampus [3–5, 51]. The increased LTP in the DG, where neurogenesis occurs, indicates the newborn neurons may have a functional role in this process. Though the rate of neurogenesis induced by exercise is low, the LTP threshold in young cells is lower and LTP lasts longer [52]. In contrast to increased LTP observed in the running rodents, LTD seems unaffected by exercising [3]. These results suggest that newborn cells have an important role in brain plasticity and their contribution can be enhanced with exercise.

Studies on rodents have shown that performance involving different spatial memory tests (Morris water maze, Y-maze, T-maze and radial arm maze) was better in exercising subjects [3]. Human studies have obtained similar results and suggested an enhancement in cognitive functions with exercise. Pereira and colleagues showed that aerobic exercise training of healthy individuals for 12 weeks significantly increased cerebral blood volume in the DG as well as cognitive functions [53]. Together, the evidences from different studies suggest that both aerobic and resistance exercise improve learning and spatial memory and reduce the cognitive loss related to ageing or neurodegenerative diseases [53–55].

3.1.1. Mediators of exercise-induced neuroplasticity

Though the exact mechanisms responsible for the increased neuroplasticity with exercise are not clear, various neurotrophic and angiogenic factors as well as neurotransmitter systems may be involved in these processes.

3.1.1.1. Trophic factors

During development of the brain and spinal cord, proliferation and differentiation of stem cells and progenitor cells are regulated by a variety of growth factors. It has been reported that growth factors are still effective in the adult period and play a significant role in synaptic plasticity, neurogenesis and cognitive functions. The increased neuroplasticity with exercise has been shown to have a contribution from growth factors such as brain-derived neurotrophic factor (BDNF), insulin-like growth factor-1 (IGF-1), vascular endothelial growth factor (VEGF), nerve growth factor (NGF) and basic fibroblast growth factor-2 (bFGF-2).

BDNF is the most important modulator of brain plasticity, showing effect through tropomyosin receptor kinase B (TrkB) receptor, inducing neuronal differentiation, proliferation and survival pathways. It is reported that BDNF as a neurotrophic factor has beneficial effects on the cognitive functions via its ability to improve synaptic plasticity, neurogenesis and LTP [3–7]. Studies of rodents showed 2–7 days of voluntary (running wheel) exercise causes a BDNF increase in different regions of the brain including the hippocampus [56]. Especially in the DG area of the hippocampus, increased BDNF levels remain high as long as exercise continues and for 2 weeks after exercise is stopped [57]. Remodelling of the hippocampus induced by exercise seems to have a critical contribution from BDNF, because the positive effect of exercise on neuroplasticity is prevented by genetic ablation or pharmacological blockading of the BDNF receptor TrkB [58, 59]. In consistent with results from animal studies, the clinical research in healthy subjects revealed an increase in the brain levels of BDNF, as detected in blood samples from the internal jugular vein following 3 months of endurance training [60]. In humans though specific changes in hippocampal BDNF levels are not well known, increased circulatory BDNF level has been reported following the exercise. Indeed, exercise-induced elevation in serum BDNF levels seems to return to baseline within 10–60 min after exercise and then decrease to a level lower than baseline [61]. In agreement with this result, a significant decrease in resting serum levels of BDNF is reported in trained subjects [62].

IGF-1 is primarily produced by the liver with the stimulation of growth hormone and modulates growth, differentiation and survival pathways of cells [3, 5, 7]. IGF-1 is accepted as a neurotrophic factor in the CNS due to its contribution to the processes of neurogenesis, differentiation, proliferation, synaptic plasticity of neurons. IGF-1 and its receptor expression have been shown in the hippocampus. Exercise increases IGF-1 level in the hippocampus in addition to plasma. Indeed, elevated IGF-1 level in the circulation is reflected in the brain, as IGF-1 can cross the blood–brain barrier via carrier proteins [63]. In sedentary individuals, injection of peripheral IGF-1 imitates the effects of exercise including increased neurogenesis and improves reduced neurogenesis related to age [64, 65]. There is a positive correlation between circulating IGF-1 levels and improved cognitive functions in healthy elderly [66]. As IGF-1-induced changes in the hippocampus are correlated with BDNF levels, it is proposed that the positive effects of IGF-1 on neuroplasticity may be mediated by BDNF [67]. Therefore, IGF-1 may have a direct and indirect contribution through increasing BDNF levels to spatial learning and memory processes.

It has been reported that adult hippocampal neurogenesis occurs near the local microvasculature and exercise stimulates endothelial cell proliferation and angiogenesis in this area of the brain. On the 3rd day of running, angiogenic processes start in addition to neurogenesis, while after a 50-day exercise period, the capillary intensity in the DG area appears to increase [49, 68]. Pereira and colleagues have shown that aerobic exercise increases blood flow in the hippocampus and improves cognitive functions in both mice and human [53]. This effect of exercise on the vascular system is reported to be mediated by IGF-1 and VEGF [3, 5, 7]. VEGF is a factor induced by hypoxia, stimulating angiogenesis mediated by tyrosine kinase receptors in endothelial cells. In addition to angiogenic effects, VEGF shows neurotrophic effect and induces growth and proliferation of progenitor cells [69]. Similar to the increase in the circulating IGF-1 levels, an increase in VEGF concentration is also reported during exercise, while peripheral IGF-1 and VEGF blocking inhibit neurogenesis induced by running [63, 69].

Other trophic factors that have been shown to be regulated by exercise and influence adult neurogenesis include bFGF-2 and NGF. However, as the increase in these factors is small and temporary, especially when compared with BDNF, they are slightly less important for the neuroplasticity induced by exercise [3].

3.1.1.2. Neurotransmitters

In addition to neurotrophic and angiogenic factors, it is known that some neurotransmitter systems are included in the regulation of neuroplasticity. For example, the most important excitatory neurotransmitter, with a key role in learning and memory processes of glutamate, stimulates neural progenitor cells and increases neurogenesis [3, 6, 7]. The expression of NR2A and NR2B subunits of the NMDA receptor increases after running exercise, and this elevation contributes to neuroplasticity induced by exercise [51, 70]. Another main neurotransmitter of the CNS with inhibitory effects, GABA, is known to affect neuroplasticity. Indeed, both inhibitor and activator effects of GABAergic system on the proliferation of progenitor cell have been reported in the hippocampus [71, 72]. During exercise, genes related to the GABAergic system ($GABA_A$ receptor, glutamate decarboxylase GAD65) are down-regulated, a system which may modify hippocampal neuroplasticity [73].

Another neurotransmitter system activated by exercise is the endocannabinoid (eCB) system. eCBs are produced by both central and peripheral tissues and during exercise circulating concentrations increase. A single bout of endurance exercise (70–80% maximal heart rate) provided optimal increase in eCB [74]. Cannabinoid receptor type 1 (CB1R) is found in different regions of the brain including frontal cortex, amygdala, hippocampus and hypothalamus [75]. Voluntary wheel-running exercise increases the agonist binding region for CB1R in the hippocampus and the levels of the endocannabinoid anandamide [76]. These alterations seem to be necessary for exercise-induced neuroplasticity in the hippocampus and are suggested to be mediators of the anti-depressant and anxiolytic effects of exercise [77].

Serotonin, dopamine and norepinephrine are the central neurotransmitters known as monoamines. Studies demonstrated that monoamines can regulate synaptic plasticity, neuronal survival and mood. It is suggested that monoamines may have a role in the exercise-induced positive effects on brain. Long-term exercise induces significant elevation of norepinephrine and serotonin levels in different brain areas, compared to the sedentary controls [78]. Klempin et al. studied tryptophan hydroxylase-2-deficient mice and investigated the neurogenic effects of serotonin deficiency [79]. The researchers found that basal neurogenesis in the hippocampus of these animals was normal; however, they showed exercise-induced neurogenesis to be reduced. These results indicate that the serotonergic system plays an important role in the effect of exercise on neuroplasticity.

3.2. Exercise and mood

Though chronic and uncontrolled stress has a negative effect on mood, it is understood that differences in the type of stress may change results [80]. For example, voluntary exercise is a stress factor (physical stress) due to activation of the sympathetic nervous system and hypothalamic-pituitary-adrenal axis [81]. However, exercise also causes release of growth hormone and various neurotrophic factors different to other stress factors such as psychological stress. Additionally, in those who exercise the transformation of the stress hormone cortisol to the inactive form of cortisone is higher and this situation is not observed with psychological stress. Finally, as exercise can be ceased whenever the person desires, it is a different to other types of stress and causes different results [7].

Clinical studies of patient with mild-to-moderate levels of depression have shown that exercise reduces depressive symptoms [82]. Additionally, animal studies showed that wheel-running activity had anti-depressant-like effects on learned helplessness, forced swim test and tail suspension tests [83]. Though the underlying mechanism responsible for the anti-depressant-like effects of exercise is not fully known, neurotrophic factors and neurotransmitter systems may provide significant contribution. For example, BDNF, IGF-1 and VEGF are thought to regulate anti-depressant-like effects of exercise [77]. Of these, BDNF may be more important than the others, because depression is related to BDNF deficiency of both the circulation and CNS. In depressive patients, peripheral BDNF levels are at lower levels compared to control [84]. Moreover, postmortem studies have shown low levels of BDNF in the cortical and hippocampal areas of the brain [85]. Studies using a variety of behaviour models have confirmed the effect of BDNF on mood by showing anti-depressant and anxiolytic effect of recombinant BDNF administration into the brain [86]. More importantly, reduced BDNF levels and depressive behaviour associated with acute and chronic stress are successfully prevented by exercise [77].

The reduction in hippocampal VEGF expression during chronic stress in rodents has shown that VEGF may be related to depression [77]. Peripheral blocking of VEGF exerts prevention of exercise-induced antidepressant-like effects, neurogenesis, as well as angiogenesis in the hippocampus [69, 87]. Additionally, VEGF polymorphism (VEGF2578C/A) causing low serum VEGF levels in humans may induce the development of treatment-resistant depression [88]. When VEGF and IGF-1 are injected into the brain, they show anti-depressant effects [89, 90]. Peripheral IGF-1 blocking prevents exercise-induced anti-depressant effects and hippocampal neurogenesis showing the importance of IGF-1 [63, 89].

As acute and chronic exercise causes an increase in plasma endorphin levels, it is proposed that endorphins may play a role in the anti-depressant effects of exercise [7]. Though endorphins are related to positive mood and well-being, the role of endorphins in exercise is still controversial as there are studies showing increased peripheral endorphins do not affect the brain [7, 91]. Monoamines are neurotransmitters with anti-depressant effects, and it is thought they may mediate the effects of exercise [92]. It has been reported that exercise increases monoamine levels in different brain area. In rodents, altered serotonin levels and serotonin metabolism has been reported in different brain area such as striatum, hippocampus, hypothalamus, cerebral cortex and brain stem following acute and chronic exercise [93]. In clinical practice, serotonergic medications are used as anti-depressants and actually exercise may be as effective as these medications for mild and moderate depression [94]. As mentioned before, eCB system may contribute to exercise-induced antidepressant-like effects [95]. Indeed, exercise-induced increase in peripheral and central concentrations of eCBs and monoamines are thought to mediate the anti-depressant and anxiolytic effects of exercise.

3.3. Exercise and ageing

Regular exercise activates neurogenesis, angiogenesis, synaptogenesis and synaptic functions in the brain via a variety of neurobiological modulators, ensuring cognitive healing. As a result, it is thought that regular exercise may prevent or improve cognitive loss in both healthy elderly people and those with risk of dementia.

In the normal ageing process, the neural plasticity of the brain reduces and this is accompanied by cognitive declines [2, 4]. The positive effects of exercise on the brain continue in elderly individuals though not as strongly as for young individuals. It has been shown that exercise exerts protective effects against cognitive decline and brain atrophy related to ageing. Kronenberg and colleagues reported that 3–9 months of exercise partially restored the reduced cell proliferation related to ageing and increased the number of mature cells [96].

Regular exercise reversed the age-related LTP decrease and it was accompanied by improved memory, increased hippocampal neurogenesis and increased BDNF levels in middle-aged rodents [97, 98]. Reduced BDNF and TrkB expression in the DG region was restored by 5 weeks [99] and 8 months [97] of forced treadmill exercise and by voluntary exercise [98] in middle-aged rats. This effect was in parallel with increased neurogenesis and improved cognitive functions [97–99]. Similarly, 8 weeks of treadmill exercise restored reduced NGF levels in the hippocampus of elderly rats [100]. Tsai et al. showed that 1 year of regular resistance exercise increased IGF-1 levels and improved cognitive functions in the elderly [55].

Human studies have identified a positive correlation between the incidence, duration and distance of daily walking activity by the elderly and the size of the hippocampus [101]. A similar study investigated the correlation between cardiovascular fitness and hippocampus volume in the elderly and showed a positive correlation between aerobic fitness (VO_2 peak) and right and left hippocampus size [102]. Elderly individuals participating in a 6-month aerobic exercise programme (60–70% of maximal heart rate, 3 times per week, 1 h) were observed to have increased grey matter and white matter volumes in the brain [103]. Cognition studies have found that cognitive functions in elderly individuals completing 3 h or more of aerobic exercise per week for at least the last 2 years were better than in sedentary individuals [104]. Another study observed that sedentary elderly individuals who participated in a 24-week aerobic exercise programme had improved cognitive functions [105]. Taken together, these results suggest that exercise is a good strategy to improve age-related brain atrophy and cognitive decline.

3.4. Exercise and neurodegenerative diseases

Neurodegenerative diseases, such as Alzheimer's disease (AD), Parkinson's disease (PD) and Huntington's disease (HD), show general properties of progressive loss of neurons. Indeed, neuronal degeneration predominantly affects different neuron groups and it is well documented that dopaminergic neurons in PD, striatal GABAergic neurons in HD and cortical and hippocampal cholinergic neurons in AD are progressively degenerated during diseases. Certain degrees of learning and memory loss are associated with progressive structural and functional deterioration of neurons in all neurodegenerative diseases [4, 6]. As the exercise stimulates neurogenesis and improves cognition in both young and old individuals, it is reasonable to think that it may have protective effect against neurodegenerative diseases.

3.4.1. Alzheimer's Disease

Pathologically, AD is characterized by acetylcholine depletion, amyloid (or senile) plaque formation, neuronal apoptosis in the cortex and hippocampus as well as brain atrophy. Additionally, decreased cognitive functions, memory loss, behavioural changes and dementia are the other characteristics of the disease [106]. In a number of AD mice models, running was observed to improve neurogenesis and cognitive functions. For example, short-term running improved cognitive functions of elderly Tg2576 mice [107], and 5-month long-term voluntary running not only prevented plaque formation in the frontal cortex and hippocampus, but also improved spatial learning in TgCRND8 D mice [108]. Similar results were obtained from APP/PS1 double transgenic AD mouse model with treadmill exercise improving learning, memory and LTP and providing amelioration of neuropathological characteristics of the disease [109]. In humans with high risk of AD, increased physical activity has been shown to reduce hippocampal atrophy [110, 111].

3.4.2. Parkinson's disease

PD develops as a result of degeneration of dopaminergic neurons with projection from the SN pars compacta to the striatum and causes severe motor symptoms such as weakened motor activity control, akinesia, tremor and rigidity. Additionally, non-motor symptoms such as cognitive reduction, olfactory dysfunction, anxiety and depression may be observed in these patients [4, 6]. In a variety of rodent PD models, it has been shown that exercise may be beneficial in improving some neuropathological and behavioural defects [4, 6]. It has been reported that 4 weeks of exercise increases the trophic factor levels and cell proliferation rate in the striatum and additionally exerts a protective effect on tyrosine hydroxylase positive neurons in the striatum and SN [112]. Clinical studies of PD patients have shown that physical activity ensured improvements in postural stability, balance and cognitive functions [113, 114]. Even more, exercise provided attenuation of depression frequently observed in both AD and PD patients [115].

3.4.3. Huntington's Disease

HD is a progressive neurodegenerative disease caused by a genetic mutation of HD gene and characterized by neuron death in the striatum mainly, but also in some regions of the cortex and hippocampus. The most significant symptoms of HD are loss of coordination of voluntary movement, bradykinesia and rigidity. Reduced cognitive capacity observed in these patients causes progressive severe dementia over time [4, 6]. Studies on different transgenic mice models such as R6/1 HD, R6/2 HD and N171-82Q HD have shown that running wheel activity does not increase BDNF level or stimulate neurogenesis in the hippocampus [116, 117]. Nevertheless, exercise delays HD symptoms and improves cognition in similar study models [118]. The cognitive improvement caused by running in HD transgenic mice does not appear to be due to increased neurogenesis, but rather is due to structural remodelling of existing neurons in the hippocampus. Human studies have provided contradictory results. For example, Altschuler reported there was no recovery of the initiation or progression of HD symptoms in a semi-professional marathon runner [119]. On the contrary, in another study of HD patients undergoing 9 months of multidisciplinary rehabilitation, including exercise, an increase in grey matter volume in the right caudate region and bilateral dorsolateral prefrontal cortex is reported in addition to significantly improved word learning and memory performance [120]. Contradictory findings suggest that more detailed investigation into the effects of exercise in HD patients is needed.

3.5. Effective type, duration and intensity of exercise

Aerobic exercise basically has positive effects on cardiovascular and metabolic functions, while resistance exercise is very beneficial for muscle strength and bone density. As mentioned above, there is much scientific proof of the positive effects of exercise on neurologic functions. Studies have shown that both exercise types have similar positive effects on neuroplasticity. Cassilhas and colleagues in a study comparing the effects of aerobic and resistance exercise found that both types of exercise produced similar improvements in learning and spatial memory after an 8-week training period. However, as aerobic exercise causes an increase in IGF-1, BDNF,

TrkB and β-CaMKII (calcium/calmodulin-dependent kinase II] in the hippocampus, resistance exercise caused an increase in peripheral and hippocampal IGF-1, IGF-1R and AKT. According to the results of this study, aerobic and resistance exercise produces similar positive effects on learning and spatial memory by using different molecular mechanisms [121].

Molteni and colleagues investigated the time-dependent changes of the gene expression in the hippocampus induced by acute and chronic (3, 7 and 28 days] voluntary running wheel activity [73]. The researchers found that BDNF expression increased with all exercise durations and reported that BDNF played a central role in exercise-induced brain plasticity. Similarly, the CaM-K signal system increased with both acute exercise and chronic exercise; however, they reported that the MAP-K/ERK system was activated by long-term exercise.

When the effect of exercise types with different intensities is examined, mild-moderate intensity exercise seems to be more beneficial to the brain, while exhaustive exercise appears to be destructive. For example, Soya and colleagues investigated the effects of acute exercise (30 min, treadmill) at different intensities on the BDNF expression in the hippocampus in rats and showed that mild running (15 m/min) increased BDNF mRNA and protein levels [122]. Running at faster speeds (20 m/min) increases serum corticosterone levels along with only BNDF mRNA levels in the hippocampus, while causing a reduction in protein levels. The researchers stated that low-intensity exercise with minimum stress levels was more beneficial to the hippocampus. Another study investigated the effect of mild and intense treadmill exercise on cognitive function after traumatic brain injury in rats, and showed that 2 weeks of mild exercise increased hippocampal BDNF expression and improved cognitive functions [123]. In another study, Li and colleagues investigated the effects of exhaustive exercise and showed that different intensities of acute exhaustive exercise cause neuronal cell apoptosis in rats. From these results, it is understood that exhaustive exercise is not beneficial to the hippocampus and may in fact have a destructive effect [124].

4. Conclusion

The beneficial effects of exercise are most likely the result of a combination of increased number and maturation of newborn cells, modifications in synaptic plasticity and spine density and enhanced angiogenesis in the hippocampus. The most pronounced changes induced by exercise are observed in the DG subfield within the hippocampus. Exercise-induced increase in hippocampal neuroplasticity appears to be mediated by a variety of neurotrophic and angiogenic factors and neurotransmitter systems. Changes in hippocampus morphology and chemistry may mediate beneficial effects of exercise, such as improved learning and memory, anti-depressant-like and anxiolytic effects, reduced cognitive decline related to ageing and delay/reduction in symptoms of neurodegenerative disease.

Although further research is needed to understand the cellular mechanism responsible for the effects of exercise on the brain, it is clear that exercise could be used to maintain and improve cognitive function throughout the lifespan. It should be also noted that exercise is the most effective, low-cost and low-tech method for successful ageing, and therefore, it has a great potential to prevent or reduce age-related cognitive decline. Neurologically to ensure the optimum benefit from exercise, regular, mild-moderate intensity of the aerobic or resistance exercise appears to be appropriate, while it is recommended to avoid exhaustive type of exercise.

Author details

Selma Cırrık[1*] and Gulay Hacioglu[2]

*Address all correspondence to: selmacrrk@yahoo.com

1 Department of Physiology, Medical Faculty, Ordu University, Ordu, Turkey

2 Department of Physiology, Medical Faculty, Giresun University, Giresun, Turkey

References

[1] Ross M.D., Malone E., Florida-James G. Vascular ageing and exercise: focus on cellular reparative processes. Oxid Med Cell Longev. 2016;2016:3583956. doi: 10.1155/2016/3583956.

[2] Brown W.M., Davison G.W., McClean C.M., Murphy M.H. A systematic review of the acute effects of exercise on immune and inflammatory indices in untrained adults. Sports Med Open. 2015;1(1):35.

[3] Vivar C., Potter M.C., van Praag H. All about running: synaptic plasticity, growth factors and adult hippocampal neurogenesis. Curr Top Behav Neurosci. 2013;15:189–210. doi: 10.1007/7854_2012_220.

[4] Yau S.Y., Gil-Mohapel J., Christie B.R., So K.F. Physical exercise-induced adult neurogenesis: a good strategy to prevent cognitive decline in neurodegenerative diseases? Biomed Res Int. 2014;2014:403120. doi:10.1155/2014/403120.

[5] Cassilhas R.C., Tufik S., de Mello M.T. Physical exercise, neuroplasticity, spatial learning and memory. Cell Mol Life Sci. 2016;73(5):975–983. doi:10.1007/s00018-015-2102-0.

[6] van Praag H. Neurogenesis and exercise: past and future directions. Neuromol Med. 2008;10(2):128–140. doi:10.1007/s12017-008-8028-z.

[7] Heijnen S., Hommel B., Kibele A., Colzato L.S. Neuromodulation of aerobic exercise-a review. Front Psychol. 2016;6:1890. doi:10.3389/fpsyg.2015.01890.

[8] Martini F.H., Nath J.L., Bartholomew E.F. Fundamentals Of Anatomy & Physiology. 9th ed. USA: Benjamin-Cummings Publishing Company; 2011. 1114 p. ISBN: 0321709330.

[9] Hendrickson R. The Facts on File Encyclopedia of Word and Phrase Origins. 3rd ed. USA: Checkmark Books; 2004. 822 p. ISBN: 0816059926.

[10] Ackerman S. Discovering the Brain. 1st ed. USA: National Academies Press; 1992. 194 p. DOI: ISBN: 0-309-57230-4.

[11] Kandel E.R., Schwartz J.H., Jessel T.M. Principles of Neural Science. 3rd ed. Norwalk, Connecticut, USA: Appleton and Lance,; 1991. 1135 p. DOI: ISBN 0-8385-8068-8.

[12] Berardi N., Sale A., Maffei L. Brain structural and functional development: genetics and experience. Dev Med Child Neurol. 2015;57:4–9. doi:10.1111/dmcn.12691.

[13] Konrad K., Firk C., Uhlhaas P.J. Brain development during adolescence: neuroscientific insights into this developmental period. Dtsch Arztebl Int. 2013;110(25):425–431.

[14] Lebel C., Beaulieu C. Longitudinal development of human brain wiring continues from childhood into adulthood. J Neurosci. 2011;31:10937–10947.

[15] Valenzuela F., Puglia M.P., Zucca S. Focus on: neurotransmitter systems. Alcohol Res Health. 2011;34:1.

[16] Catania M.V., D'Antoni S., Bonaccorso C.M., Aronica E., Bear M.F., Nicoletti F. Group I metabotropic glutamate receptors: a role in neurodevelopmental disorders? Mol Neurobiol. 2007;35:298–307.

[17] Calero M.D., Navarro E. Cognitive plasticity as a modulating variable on the effects of memory training in elderly persons. Arch Clin Neuropsychol. 2007;22:63–72.

[18] Rabipour S., Raz A. Training the brain: fact and fad in cognitive and behavioral remediation. Brain Cogn. 2012;79:159–179.

[19] Lustig C., Shah P., Seidler R., Reuter-Lorenz P.A. Aging, training, and the brain: a review and future directions. Neuropsychol Rev. 2009;19(4):504–22. doi:10.1007/s11065-009-9119-9.

[20] Gomez-Pinilla F., Hillman C. The influence of exercise on cognitive abilities.. Compr Physiol. 2013;3(1):403–428. doi:10.1002/cphy.c110063.

[21] Ganguly K., Poo M.M. Activity-dependent neural plasticity from bench to bedside. Neuron. 2013;80(3):729–41.

[22] Fuchs E., Flügge G. Adult neuroplasticity: more than 40 years of research. Neural Plast. 2014;2014:541870. doi:10.1155/2014/541870.

[23] Zilles K. Neuronal plasticity as an adaptive property of the central nervous system. Ann Anat. 1992;174(5):383–391.

[24] Stiles J. Neural plasticity and cognitive development. Dev Neuropsychol. 2000;18(2): 237–272.

[25] Wan C.Y., Schlaug G. Music making as a tool for promoting brain plasticity across the life span. Neuroscientist. 2010;16(5):566–577. doi:10.1177/1073858410377805.

[26] Hübener M., Bonhoeffer T. Neuronal plasticity: beyond the critical period. Cell. 2014;159(4):727–737.

[27] Bliss T.V.P., Lomo T. Long lasting potentiation of synaptic transmission in the dentate area of the anaesthetized rabbit following stimulation of the perforant path. J Physiol. 1973;232(2):331–356.

[28] Zhao M., Momma S., Delfani K., Carlen M., Cassidy R.M., Johansson C.B., et al. Evidence for neurogenesis in the adult mammalian substantia nigra. Proc Natl Acad Sci U S A. 2003;100(13):7925–7930.

[29] Sawada M., Sawamoto K. Mechanisms of neurogenesis in the normal and injured adult brain. Keio J Med. 2013;62:13–28.

[30] Voloboueva L.A., Giffard R.G. Inflammation, mitochondria, and the inhibition of adult neurogenesis. J Neurosci Res. 2011;89(12):1989–1996.

[31] Monje M.L., Toda H., Palmer T.D. Inflammatory blockade restores adult hippocampal neurogenesis. Science. 2003;302(5651):1760–1765.

[32] Luine V., Frankfurt M. Interactions between estradiol, BDNF and dendritic spines in promoting memory. Neuroscience. 2013;239:34–45. doi:10.1016/j.neuroscience. 2012.10.019.

[33] Hjorth-Simonsen A. Some intrinsic connections of the hippocampus in the rat: an experimental analysis. J Comp Neurol. 1973;147(2):145–61. doi:10.1002/cne.901470202.

[34] Laurberg S. Commissural and intrinsic connections of the rat hippocampus. J Comp Neurol. 1979;184(4):685–708. doi:10.1002/cne.901840405.

[35] Bliss T.V., Collingridge G.L. A synaptic model of memory: long-term potentiation in the hippocampus. Nature. 1993;361(6407):31–39.

[36] Bliss T., Lømo T. Long-lasting potentiation of synaptic transmission in the dentate area of the anaesthetized rabbit following stimulation of the perforant path. J Physiol. 1973;232(2):331–356.

[37] Gureviciene I., Ikonen S., Gurevicius K., Sarkaki A., van Groen T., Pussinen R., et al. Normal induction but accelerated decay of LTP in APP + PS1 transgenic mice. Neurobiol Dis. 2004;15:188–195.

[38] Collingridge G.L., Kehl S.J., McLennan H. Excitatory amino acids in synaptic transmission in the Schaffer collateral-commissural pathway of the rat hippocampus. J Physiol. 1983;334:33–46.

[39] Croucher M.J., Collins J.F., Meldrum B.S. Anticonvulsant action of excitatory amino acid antagonists. Science. 1982;216:899–901.

[40] Simon R.P., Swan J.H., Griffiths T., Meldrum B.S. Blockade of N-methyl-D-aspartate receptors may protect against ischemic damage in the brain. Science. 1984;226(4676): 850–852.

[41] Davies S.N., Lodge D. Evidence for involvement of N-methylaspartate receptors in 'wind-up' of class 2 neurones in the dorsal horn of the rat. Brain Res. 1987;424:402–406.

[42] Vorhees C., Williams M.T. Assessing spatial learning and memory in rodents. ILAR J. 2014;55(2):310–332.

[43] Morris R.G.M. Spatial localization does not require the presence of local cues. Learn Motiv. 1981;12:239–260.

[44] DeVriesa A.C, Nelsona R.J., Traystmanb R.J., Hurn P.D. Cognitive and behavioral assessment in experimental stroke research: will it prove useful? Neurosci Biobehav Rev. 2001;25(4):325–342.

[45] Olton D.S., Samuelson R.J. Remembrance of placed passed: spatial memory in rats. J Exp Psychol: Anim Behav Process. 1976;2:97–116.

[46] Olton D.S., Collison C., Werz M.A. Spatial memory and radial arm maze performance of rats. Learn Motiv. 1977;8(3):289–314. doi:10.1016/0023-9690(77)90054-6.

[47] Brown M.F., Guimetti G.W. Spatial pattern learning in the radial arm maze. Learning Behav. 2006;34(1):102–108. doi:10.3758/BF03192875.

[48] Zhao C., Teng E.M., Summers R.G., Jr, Ming G.L., Gage F.H. Distinct morphological stages of dentate granule neuron maturation in the adult mouse hippocampus. J Neurosci. 2006;26(1):3–11.

[49] Van der Borght K., Kóbor-Nyakas D.E., Klauke K., Eggen B.J., Nyakas C., Van der Zee E.A., et al. Physical exercise leads to rapid adaptations in hippocampal vasculature: temporal dynamics and relationship to cell proliferation and neurogenesis. Hippocampus. 2009;19(10):928–936. doi:10.1002/hipo.20545.

[50] Fernandes J., Kramer Soares J.C., Zaccaro do Amaral Baliego L.G., Arida R.M. A single bout of resistance exercise improves memory consolidation and increases the expression of synaptic proteins in the hippocampus. Hippocampus. 2016;26(8):1096-1103. doi: 10.1002/hipo.22590.

[51] Vasuta C., Caunt C., James R., Samadi S., Schibuk E., Kannangara T., et al. Effects of exercise on NMDA receptor subunit contributions to bidirectional synaptic plasticity in the mouse dentate gyrus. Hippocampus. 2007;17(12):1201–1208.

[52] Schmidt-Hieber C., Jonas P., Bischofberger J. Enhanced synaptic plasticity in newly generated granule cells of the adult hippocampus. Nature. 2004;429(6988):184–187.

[53] Pereira A.C., Huddleston D.E., Brickman A.M., Sosunov A.A., Hen R., McKhann G.M., et al. An *in vivo* correlate of exercise-induced neurogenesis in the adult dentate gyrus. Proc Natl Acad Sci U S A. 2007;104(13):5638–5643.

[54] Colcombe S., Kramer A.F. Fitness effects on the cognitive function of older adults: a meta-analytic study. Psychol Sci. 2003;14(2):125–130.

[55] Tsai C.L., Wang C.H., Pan C.Y., Chen F.C. The effects of long-term resistance exercise on the relationship between neurocognitive performance and GH, IGF-1, and homocysteine levels in the elderly. Front Behav Neurosci. 2015;9:23. doi:10.3389/fnbeh.2015.00023.

[56] Berchtold N.C., Chinn G., Chou M., Kesslak J.P., Cotman C.W. Exercise primes a molecular memory for brain-derived neurotrophic factor protein induction in the rat hippocampus. Neuroscience. 2005;133(3):853–861.

[57] Berchtold N.C., Castello N., Cotman C.W. Exercise and time-dependent benefits to learning and memory. Neuroscience. 2010;167(3):588–597.

[58] Li Y., Luikart B.W., Birnbaum S., Chen J., Kwon C.H., Kernie S.G., Bassel-Duby R., Parada L.F. TrkB regulates hippocampal neurogenesis and governs sensitivity to antidepressive treatment. Neuron. 2008;59(3):399–412. doi:10.1016/j.neuron.2008.06.023.

[59] Vaynman S., Ying Z., Gomez-Pinilla F. Hippocampal BDNF mediates the efficacy of exercise on synaptic plasticity and cognition. Eur J Neurosci. 2004;20(10):2580–2590.

[60] Seifert T., Brassard P., Wissenberg M., Rasmussen P., Nordby P., Stallknecht B. et al. Endurance training enhances BDNF release from the human brain. Am J Physiol. 2010;298(2):R372–R377. doi:10.1152/ajpregu.00525.2009.

[61] Knaepen K., Goekint M., Heyman E.M., Meeusen R. Neuroplasticity exercise-induced response of peripheral brain-derived neurotrophic factor: a systematic review of experimental studies in human subjects. Sports Med. 2010;40(9):765–801. doi:10.2165/11534530-000000000-00000.

[62] Currie J., Ramsbottom R., Ludlow H., Nevill A., Gilder M. Cardio-respiratory fitness, habitual physical activity and serum brain derived neurotrophic factor (BDNF) in men and women. Neurosci Lett. 2009;451(2):152–155. doi:10.1016/j.neulet.2008.12.043.

[63] Trejo J.L., Carro E., Torres-Aleman I. Circulating insulin-like growth factor I mediates exercise-induced increases in the number of new neurons in the adult hippocampus. J Neurosci. 2001;21(5):1628–1634.

[64] Aberg M.A., Aberg N.D., Hedbäcker H., Oscarsson J., Eriksson P.S. Peripheral infusion of IGF-I selectively induces neurogenesis in the adult rat hippocampus. J Neurosci. 2000;20(8):2896–2903.

[65] Lichtenwalner R.J., Forbes M.E., Bennett S.A., Lynch C.D., Sonntag W.E., Riddle D.R. Intracerebroventricular infusion of insulin-like growth factor-I ameliorates the age-related decline in hippocampal neurogenesis. Neuroscience. 2001;107(4):603–613.

[66] Arwert L.I., Deijen J.B., Drent M.L. The relation between insulin-like growth factor I levels and cognition in healthy elderly: a meta-analysis. Growth Hormone IGF Res. 2005;15(6):416–422.

[67] Ding Q., Vaynman S., Akhavan M., Ying Z., Gomez-Pinilla F. Insulin-like growth factor I interfaces with brain-derived neurotrophic factor-mediated synaptic plasticity to modulate aspects of exercise-induced cognitive function. Neuroscience. 2006;140(3): 823–833.

[68] Clark P.J., Brzezinska W.J., Puchalski E.K., Krone D.A., Rhodes J.S. Functional analysis of neurovascular adaptations to exercise in the dentate gyrus of young adult mice associated with cognitive gain. Hippocampus. 2009;19(10):937–950. doi:10.1002/hipo.20543.

[69] Fabel K., Fabel K., Tam B., Kaufer D., Baiker A., Simmons N., et al. VEGF is necessary for exercise-induced adult hippocampal neurogenesis. Eur J Neurosci. 2003;18(10): 2803–2812.

[70] Lou S.J., Liu J.Y., Chang H., Chen P.J. Hippocampal neurogenesis and gene expression depend on exercise intensity in juvenile rats. Brain Res. 2008;1210:48–55. doi:10.1016/j.brainres.2008.02.080.

[71] Giachino C., Barz M., Tchorz J.S., Tome M., Gassmann M., Bischofberger J., et al. GABA suppresses neurogenesis in the adult hippocampus through GABAB receptors. Development. 2014;141(1):83–90. doi:10.1242/dev.102608.

[72] Tozuka Y., Fukuda S., Namba T., Seki T., Hisatsune T. GABAergic excitation promotes neuronal differentiation in adult hippocampal progenitor cells. Neuron. 2005;47(6): 803–815.

[73] Molteni R., Ying Z., Gómez-Pinilla F. Differential effects of acute and chronic exercise on plasticity-related genes in the rat hippocampus revealed by microarray. Eur J Neurosci. 2002;16:1107–1116.

[74] Raichlen D.A., Foster A.D., Seillier A., Giuffrida A., Gerdeman G.L. Exercise-induced endocannabinoid signaling is modulated by intensity. Eur J Appl Physiol. 2013;113(4): 869–875. doi:10.1007/s00421-012-2495-5.

[75] Marco E.M., García-Gutiérrez M.S., Bermúdez-Silva F.J., Moreira F.A., Guimarães F., Manzanares J., et al. Endocannabinoid system and psychiatry: in search of a neurobiological basis for detrimental and potential therapeutic effects. Front Behav Neurosci. 2011;5(63): 1–23. doi:10.3389/fnbeh.2011.00063.

[76] Hill M.N., Titterness A.K., Morrish A.C., Carrier E.J., Lee T.T., Gil-Mohapel J., et al. Endogenous cannabinoid signaling is required for voluntary exercise-induced en-

hancement of progenitor cell proliferation in the hippocampus. Hippocampus. 2010;20(4):513–523. doi:10.1002/hipo.20647.

[77] Ota K.T., Duman R.S. Environmental and pharmacological modulations of cellular plasticity: role in the pathophysiology and treatment of depression. Neurobiol Dis. 2013;57:28–37. doi:10.1016/j.nbd.2012.05.022.

[78] Ma Q. Beneficial effects of moderate voluntary physical exercise and its biological mechanisms on brain health. Neurosci Bull. 2008;24(4):265–270. doi:10.1007/s12264-008-0402-1.

[79] Klempin F., Beis D., Mosienko V., Kempermann G., Bader M., Alenina N. Serotonin is required for exercise-induced adult hippocampal neurogenesis. J Neurosci. 2013;33(19):8270–8575. doi:10.1523/JNEUROSCI.5855-12.2013.

[80] Ranabir S., Reetu K. Stress and hormones. Indian J Endocrinol Metab. 2011;15(1):18–22. doi:10.4103/2230-8210.77573.

[81] Labsy Z., Prieur F., Le Panse B., Do M.C., Gagey O., Lasne F., et al. The diurnal patterns of cortisol and dehydroepiandrosterone in relation to intense aerobic exercise in recreationally trained soccer players. Stress. 2013;16(2):261–265. doi:10.3109/10253890.2012.707259.

[82] Herring M.P., Puetz T.W., O'Connor P.J., Dishman R.K. Effect of exercise training on depressive symptoms among patients with a chronic illness: a systematic review and meta-analysis of randomized controlled trials. Arch Intern Med. 2012;172(2):101–111. doi:10.1001/archinternmed.2011.696.

[83] Duman C.H., Schlesinger L., Russell D.S., Duman R.S. Voluntary exercise produces antidepressant and anxiolytic behavioral effects in mice. Brain Res. 2008;1199:148–58. doi:10.1016/j.brainres.2007.12.047.

[84] Bocchio-Chiavetto L., Bagnardi V., Zanardini R., Molteni R., Nielsen M.G., Placentino A., et al. Serum and plasma BDNF levels in major depression: a replication study and meta-analyses. World J Biol Psychiatry. 2010;11(6):763–773. doi:10.3109/15622971003611319.

[85] Karege F., Bondolfi G., Gervasoni N., Schwald M., Aubry J.M., Bertschy G. Low brain-derived neurotrophic factor (BDNF) levels in serum of depressed patients probably results from lowered platelet BDNF release unrelated to platelet reactivity. Biol Psychiatry. 2005;57(9):1068–1072.

[86] Schmidt H.D., Duman R.S. Peripheral BDNF produces antidepressant-like effects in cellular and behavioral models. Neuropsychopharmacology. 2010;35(12):2378–2391. doi:10.1038/npp.2010.114.

[87] Kiuchi T., Lee H., Mikami T. Regular exercise cures depression-like behavior via VEGF-Flk-1 signaling in chronically stressed mice. Neuroscience. 2012;207:208–127. doi:10.1016/j.neuroscience.2012.01.023.

[88] Viikki M., Anttila S., Kampman O., Illi A., Huuhka M., Setälä-Soikkeli E., et al. Vascular endothelial growth factor (VEGF) polymorphism is associated with treatment resistant depression. Neurosci Lett. 2010;477(3):105–108. doi:10.1016/j.neulet.2010.04.039.

[89] Duman C.H., Schlesinger L., Terwilliger R., Russell D.S., Newton S.S., Duman R.S. Peripheral insulin-like growth factor-I produces antidepressant-like behavior and contributes to the effect of exercise. Behav Brain Res. 2009;198(2):366–371. doi:10.1016/j.bbr.2008.11.016.

[90] Warner-Schmidt J.L., Duman R.S. VEGF is an essential mediator of the neurogenic and behavioral actions of antidepressants. Proc Natl Acad Sci U S A. 2007;104(11):4647–4652.

[91] Sparling P.B., Giuffrida A., Piomelli D., Rosskopf L., Dietrich A. Exercise activates the endocannabinoid system. Neuroreport. 2003;14(17):2209–2211.

[92] Lawlor D.A., Hopker S.W. The effectiveness of exercise as an intervention in the management of depression: systematic review and meta-regression analysis of randomised controlled trials. BMJ. 2001;322(7289):763–767.

[93] Dey S., Singh R.H., Dey P.K. Exercise training: significance of regional alterations in serotonin metabolism of rat brain in relation to antidepressant effect of exercise. Physiol Behav. 1992;52(6):1095–1099.

[94] Ernst C., Olson A.K., Pinel J.P., Lam R.W., Christie B.R. Antidepressant effects of exercise: evidence for an adult-neurogenesis hypothesis? J Psychiatr Neurosci. 2006;31(2):84–92.

[95] Gorzalka B.B., Hill M.N. Putative role of endocannabinoid signaling in the etiology of depression and actions of antidepressants. Prog Neuropsychopharmacol Biol Psychiatry. 2011;35(7):1575–85. doi:10.1016/j.pnpbp.2010.11.021.

[96] Kronenberg G., Bick-Sander A., Bunk E., Wolf C., Ehninger D., Kempermann G. Physical exercise prevents age-related decline in precursor cell activity in the mouse dentate gyrus. Neurobiol Aging. 2006;27(10):1505–1513.

[97] O'Callaghan R.M., Griffin E.W., Kelly A.M. Long-term treadmill exposure protects against age-related neurodegenerative change in the rat hippocampus. Hippocampus. 2009;19(10):1019–10129. doi:10.1002/hipo.20591.

[98] Marlatt M.W., Potter M.C., Lucassen P.J., van Praag H. Running throughout middle-age improves memory function, hippocampal neurogenesis and BDNF levels in female C57Bl/6 J mice. Dev Neurobiol. 2012;72(6):943–952. doi:10.1002/dneu.22009.

[99] Wu C.W., Chang Y.T., Yu L., Chen H.I., Jen C.J., Wu S.Y., et al. Exercise enhances the proliferation of neural stem cells and neurite growth and survival of neuronal progenitor cells in dentate gyrus of middle-aged mice. J Appl Physiol. 2008;105(5):1585–1594. doi:10.1152/japplphysiol.90775.2008.

[100] Chae C.H., Kim H.T. Forced, moderate-intensity treadmill exercise suppresses apoptosis by increasing the level of NGF and stimulating phosphatidylinositol 3-kinase signaling in the hippocampus of induced aging rats. Neurochem Int. 2009;55(4):208–213. doi:10.1016/j.neuint.2009.02.024.

[101] Varma V.R., Chuang Y.F., Harris G.C., Tan E.J., Carlson M.C. Low-intensity daily walking activity is associated with hippocampal volume in older adults. Hippocampus. 2015;25(5):605–615. doi:10.1002/hipo.22397.

[102] Erickson K.I., Prakash R.S., Voss M.W., Chaddock L., Hu L., Morris K.S., et al. Aerobic fitness is associated with hippocampal volume in elderly humans. Hippocampus. 2009;19(10):1030–1039. doi:10.1002/hipo.20547.

[103] Colcombe S.J., Erickson K.I., Scalf P.E., Kim J.S., Prakash R., McAuley E., et al. Aerobic exercise training increases brain volume in aging humans. J Gerontol A Biol Sci Med Sci. 2006;61(1):1166–1170.

[104] Sánchez-Horcajo R., Llamas-Alonso J., Cimadevilla J.M. Practice of aerobic sports is associated with better spatial memory in adults and older men. Exp Aging Res. 2015;41(2):193–203. doi:10.1080/0361073X.2015.1001656.

[105] Tarazona-Santabalbina F.J., Gómez-Cabrera M.C., Pérez-Ros P., Martínez-Arnau F.M., Cabo H., Tsaparas K., et al. A multicomponent exercise intervention that reverses frailty and improves cognition, emotion, and social networking in the community-dwelling frail elderly: a randomized clinical trial. J Am Med Dir Assoc. 2016;17(5):426–433. doi: 10.1016/j.jamda.2016.01.019.

[106] Hardy J., Selkoe D.J. The amyloid hypothesis of Alzheimer's disease: progress and problems on the road to therapeutics. Science. 2002;297(5580):353–356.

[107] Parachikova A., Nichol K.E., Cotman C.W. Short-term exercise in aged Tg2576 mice alters neuroinflammation and improves cognition. Neurobiol Dis. 2008;30(1):121–129. doi:10.1016/j.nbd.2007.12.008

[108] Adlard P.A., Perreau V.M., Pop V., Cotman C.W. Voluntary exercise decreases amyloid load in a transgenic model of Alzheimer's disease. J Neurosci. 2005;25(7):4217–4221.

[109] Liu H.L., Zhao G., Zhang H., Shi L.D. Long-term treadmill exercise inhibits the progression of Alzheimer's disease-like neuropathology in the hippocampus of APP/PS1 transgenic mice. Behav Brain Res. 2013;256:261–272. doi:10.1016/j.bbr.2013.08.008.

[110] Smith J.C., Nielson K.A., Woodard J.L., Seidenberg M., Durgerian S., Hazlett K.E., et al. Physical activity reduces hippocampal atrophy in elders at genetic risk for Alzheimer's disease. Front Aging Neurosci. 2014;6:61. doi:10.3389/fnagi.2014.00061.

[111] Buchman A.S., Boyle P.A., Yu L., Shah R.C., Wilson R.S., Bennett D.A. Total daily physical activity and the risk of AD and cognitive decline in older adults. Neurology. 2012;78(17):1323–1329. doi:10.1212/WNL.0b013e3182535d35.

[112] Tajiri N., Yasuhara T., Shingo T., Kondo A., Yuan W., Kadota T., et al. Exercise exerts neuroprotective effects on Parkinson's disease model of rats. Brain Res. 2010;1310:200–207. doi:10.1016/j.brainres.2009.10.075.

[113] Fisher B.E., Wu A.D., Salem G.J., Song J., Lin C.H., Yip J., et al. The effect of exercise training in improving motor performance and corticomotor excitability in people with early Parkinson's disease. Arch Phys Med Rehabil. 2008;89(7):1221–1229. doi:10.1016/j.apmr.2008.01.013.

[114] Petzinger G.M., Fisher B.E., McEwen S., Beeler J.A., Walsh J.P., Jakowec M.W. Exercise-enhanced neuroplasticity targeting motor and cognitive circuitry in Parkinson's disease. Lancet Neurol. 2013;12(7):716–726. doi:10.1016/S1474-4422(13)70123-6.

[115] Peluso M.A., Guerra de Andrade L.H. Physical activity and mental health: the association between exercise and mood. Clinics (Sao Paulo). 2005;60(1):61–70.

[116] Ransome M.I., Hannan A.J. Impaired basal and running-induced hippocampal neurogenesis coincides with reduced Akt signaling in adult R6/1 HD mice. Mol Cell Neurosci. 2013;54:93–107. doi:10.1016/j.mcn.2013.01.005.

[117] Kohl Z., Kandasamy M., Winner B., Aigner R., Gross C., Couillard-Despres S., et al. Physical activity fails to rescue hippocampal neurogenesis deficits in the R6/2 mouse model of Huntington's disease. Brain Res. 2007;1155:24–33.

[118] Pang T.Y., Stam N.C., Nithianantharajah J., Howard M.L., Hannan A.J. Differential effects of voluntary physical exercise on behavioral and brain-derived neurotrophic factor expression deficits in Huntington's disease transgenic mice. Neuroscience. 2006;141(2):569–584.

[119] Altschuler E.L. Strenuous, intensive, long-term exercise does not prevent or delay the onset of Huntington's disease. Med Hypotheses. 2006;67(6):1429–1430.

[120] Cruickshank T.M., Thompson J.A., Domínguez D.J.F., Reyes A.P., Bynevelt M., Georgiou-Karistianis N., et al. The effect of multidisciplinary rehabilitation on brain structure and cognition in Huntington's disease: an exploratory study. Brain Behav. 2015;5(2):e00312. doi:10.1002/brb3.312.

[121] Cassilhas R.C., Lee K.S., Fernandes J., Oliveira M.G., Tufik S., Meeusen R., et al. Spatial memory is improved by aerobic and resistance exercise through divergent molecular mechanisms. Neuroscience. 2012;202:309–317. doi:10.1016/j.neuroscience.2011.11.029.

[122] Soya H., Nakamura T., Deocaris C.C., Kimpara A., Iimura M., Fujikawa T., et al. BDNF induction with mild exercise in the rat hippocampus. Biochem Biophys Res Commun. 2007;358(4):961–967.

[123] Shen X., Li A., Zhang Y., Dong X., Shan T., Wu Y., et al. The effect of different intensities of treadmill exercise on cognitive function deficit following a severe controlled cortical impact in rats. Int J Mol Sci. 2013;14(11):21598–21612. doi: 10.3390/ijms141121598.

[124] Li S., Liu J., Yan H. Medium-intensity acute exhaustive exercise induces neural cell apoptosis in the rat hippocampus. Neural Regen Res. 2013;8(2):127–132. doi:10.3969/j.issn.1673-5374.2013.02.004.

Chapter 5

Exercise for Hypertension

Chulvi-Medrano Iván, Sanchis-Cervera José,
Tortosa-Martínez Juan and
Cortell-Tormo Juan Manuel

Additional information is available at the end of the chapter

http://dx.doi.org/10.5772/65035

Abstract

High blood pressure (HBP) does not cause discomfort but still, you need to take care of and treat it. Otherwise, over time, it can damage the heart, kidneys, brain, and eyes. Lifestyle changes are essential in HBP, and physical activity is a parameter of great influence. In order to achieve the benefits derived from physical activity, it must be adequately prescribed, an aspect that will be developed in this chapter. The first section addresses the physiopathology of hypertension, with special interest in the pathological mechanisms that may induce hypertension, devices for monitoring blood pressure (BP), and an overview of the particularities that present hypertension in the presence of other pathologies and over the life span. Second section focuses on exercise prescription for hypertensive people, exploring each type of exercise that has been proved to be effective. We discuss for each type of exercise, the benefits, mechanisms involved in these benefits, the appropriate dose of exercise, and other methodological considerations including risk management issues. We conclude with a clinical case study. A detailed exercise training program will be developed for this particular case study in order to try to bridge the gap between theory and practice.

Keywords: primary hypertension, blood pressure, resistance training, aerobic exercise, lifestyle

1. Introduction

High blood pressure (HBP) is the greatest public health problem worldwide because of its high prevalence and because it is associated with increased cardiovascular and renal complications. In the year 2000, hypertension in the adult population stood about 26.4%. Projected calculations for 2025 estimate the prevalence at 29.0%. In absolute terms worldwide, this increase will go from 972 million hypertensives in 2000 to 1,560,000,000 in 2025, representing a raise of approximately 60% [1].

Hypertension is the most prevalent cardiovascular disease (CVD), affecting 20–50% of the population in developed countries. Its prevalence increases with age, especially after 50 years old when it affects more than 50% of the population. All segments of the population (pregnant women, elderly, people with diabetes, etc.) are susceptible to developing a pathological elevation of blood pressure (BP). Hypertension is usually diagnosed when systolic blood pressure (SBP) is ≥140 mmHg and/or diastolic blood pressure (DBP) ≥90 mmHg, sustained constantly. However, the dividing line between normal blood pressure and hypertension is arbitrary and actually artificial. The value of BP is the result of the interaction of genetic and environmental factors (e.g., *polygenic inheritance, salt intake, smoking, stress, diet, etc.*).

A high BP has been identified as a risk factor for coronary artery disease (CAD), heart failure (HF), stroke, and peripheral arterial insufficiency. Hypertension is associated with earlier changes in organ systems, such as left ventricular (LV) hypertrophy, renal failure, retinopathy, and vascular dementia, which are grouped under the term "target organ damage" (TOD). Furthermore, hypertensive patients have a higher incidence of cardiovascular comorbidities (e.g., *diabetes, insulin resistance, dyslipidemia, etc.*) compared to the general population.

Currently, HBP represents a major challenge to public health as it is frequently associated with sudden death due to the silent nature of the condition. The first step is to secure a firm diagnosis. The European League of Hypertension recommends the use of an ambulatory BP measurement (monitoring ambulatory blood pressure (MAPA)), which is a device for automated blood pressure measurement repeatedly for 24 h. This monitoring test provides a more accurate picture of the circadian pattern of blood pressure changes over an average day and night. Recently, there has been a marked interest in this circadian pattern and the variability of BP because having an abnormal BP pattern has been related to increased cardiovascular risk and other health problems. In particular, night blood pressure seems to be the best predictor of cardiovascular events and the degree of development of cardiovascular disease. The MAPA can also differentiate permanent hypertension and hypertension secondary related to stress (white-coat effect). Scientific evidence regarding the effectiveness of the MAPA has mainly emerged from the CARDIORISC project in Spain, which includes the largest global database available with more than 40,000 patients [2].

Inadequate treatment of hypertension increases drastically the potential damage on the main target organs (*e.g., heart, brain, kidney, retina, etc.*). There are several nonpharmacological strategies for preventing and treating hypertension, such as the DASH diet and physical activity. Recent evidence suggests that higher physical activity levels are associated not only with lower diurnal BP levels but also with lower levels of night BP, which represents a healthier circadian pattern and thus a reduced cardiovascular comorbidity. There is a considerable amount of scientific evidence, including meta-analysis, supporting the antihypertensive effects of different types of exercise for blood pressure, such as continuous aerobic training. However, the appropriate type, intensity, time, and frequency of exercise remain unclear. In parallel, other types of exercise such as isometric strength training or high-intensity aerobic training (HIT) are gaining attention in the literature, showing very promising results. In order to achieve

http://dx.doi.org/10.5772/65035

and maximize the benefits of exercise safely, professionals should have updated guidelines for prescribing exercise for hypertension. These guidelines should be based on scientific evidence and provide precise practical information about how to design, implement, and control a training program.

2. Pathological mechanisms

2.1. Etiological factors

2.1.1. Genetics

From a clinical standpoint, the influence of genetics in hypertension is determined by a familial aggregation, so that the prevalence increases among relatives of first degree. Besides this clinical observation, knowledge of the genes involved in the development of hypertension is scarce. It seems clear that genetic determinants can be modified by environmental ones, so that the blood pressure or the resulting phenotype depends on the interaction of both factors. Genotypic and phenotypic research will be of great importance in the future as it may be able to help develop personalized treatments and prevention programs based on these findings [3–6].

2.1.2. Environmental factors

The main factors related to the development of hypertension are linked to the progress and changes in lifestyle and diet. The mechanisms by which these environmental determinants increase blood pressure and promote the development of cardiovascular disease are not fully clarified but it seems that a high caloric intake and decreased energy expenditure produce sympathetic hyperactivity [3–6].

2.1.3. Possible pathological mechanisms

1. *Sympathetic overactivity*. It is characterized by an imbalance between sympathetic and parasympathetic activity. Sympathetic activation may be caused by a direct stimulus by chronic stress, high caloric intake, obesity, and physical inactivity among others. 2. *Renin-angiotensin system (RAS)*. The renin-angiotensin system is undoubtedly mainly responsible for the development of vascular disease and one of the main focuses of therapeutic care. Angiotensin II is the main effector system with specific receptors at different levels that promote vasoconstriction and fluid retention, the latter thanks to the stimulation of adrenal aldosterone secretion as well as inflammatory, proaggregatory, and prothrombotic reactions. 3. *Dysfunction and endothelial injury*. The endothelium plays a fundamental role in vascular pathophysiology alterations observed in hypertension. Cardiovascular complications include both dysfunction and damage in the endothelial cell layer. In the latter case, the most plausible hypothesis is the inability of predisposed individuals to repair endothelial cell damage that occurs in normal circumstances, mainly due to a decrease in endothelial progenitor cells, primarily responsible for these repair processes. Regarding functional alterations, these consist of an imbalance

between the production of vasodilator and anti-inflammatory substances including nitric oxide and the production of pro-inflammatory and primarily endothelin vasoconstrictors. 4. *Structural changes in the arteries*. Three types of changes in arteries are described in hypertension: the presence of capillary rarefaction, hypertrophy of the middle layer of resistance arteries, and stiffness of large arteries [3–6].

3. Blood pressure monitoring (*random vs MAPA*)

The diagnosis of hypertension and the therapeutic decisions derived from this diagnosis require the highest reliability possible. The difficulties in measuring the blood pressure of a subject are derived from its variability, associated mainly with his/her physical and mental activity, limitations in the accuracy of an indirect measure (with the observer as a major source of inaccuracy), and sometimes the alert reaction produced in a person to blood pressure monitoring, which in some cases can be of clinical relevance (white-coat phenomenon).

Measuring blood pressure in a clinical setting: the reference technique of measuring blood pressure has been classically performed in the clinical setting by the auscultatory technique with the mercury sphygmomanometer, and most recently by semiautomatic and validated self-measurement devices. Monitoring ambulatory blood pressure provides real value of great importance. MAPA keeps 24-h Holter records of the blood pressure of a person. The patient may be classified with respect to the values of blood pressure in four exclusive conditions: real normotensive, sustained hypertension, white-coat hypertension, and masked hypertension. Because of its advantages, MAPA is increasingly considered a necessary element in the basic assessment of hypertensive patients, with the recommendation of some clinical guidelines to use it for confirmation of diagnosis, previous to making therapeutic decisions [3–6].

4. Morbidity of hypertension

The main clinical guidelines for the diagnosis and treatment of hypertension advise for the evaluation, to the possible extent, of the existence of subclinical organ damage. The reason for this is the consideration of organ damage as an intermediate stage in the development of cardiovascular disease, with importance in the stratification of absolute cardiovascular risk, and therefore the need for a different therapeutic approach. Multiple components of subclinical organ damage in heart, renal, and cerebral level have been described [3–6].

4.1. Main consequences

In the **heart**, hypertension causes left ventricular hypertrophy, which is the first step for developing hypertensive heart disease.

The silent damage produced in the **kidney** should be assessed in all hypertensive patients at diagnosis. Furthermore, at least annually, the evaluation should proceed by determining serum creatinine, renal blood flow, as well as the microalbuminuria.

The evaluation of **vascular injury** is more complex and its relationship with prognosis is less evident. In particular, in regard to the modifications induced in the treatment, there is no universal recommendation for vascular assessment in hypertensive patients although there is evidence about the usefulness of the ankle arm index. The ankle arm index is an indicator of the existence of peripheral arterial disease and can be altered in apparently asymptomatic patients.

5. Drugs and pharmacology associated

The aim of an antihypertensive therapy is twofold. On one hand, reducing mortality and cardiovascular morbidity associated with increased pressure and, on the other, avoiding the progression and regression of subclinical visceral organ damage. Treatment should be initiated immediately in all patients with hypertension grade III (SBP of >180; DBP of >110), regardless of the presence of other conditions, and for any other hypertensive patients in the presence of diabetes, target organ damage, or chronic kidney disease established by cardiovascular disease. Although there is no clear consensus in the clinical guidelines, there is a general indication for starting treatment in elderly individuals at values above 150 mmHg 90 [3–6].

5.1. Diuretics

Diuretics are first-line drugs, and there is more evidence available of its protective capacity than with other drugs. There are three different subgroups of diuretics: thiazide, loop diuretics of Henle, and potassium-sparing, although only the former has clearly demonstrated a clear benefit in cardiovascular pressure in hypertensive patients.

5.2. Calcium antagonists

There are three main groups of calcium-channel blockers. Their mechanism of action consists of the inhibition of dependent calcium channels in the membrane potential blocking the entry of calcium into the cell. The biggest drawback of these drugs is the frequent occurrence of side effects. In nearly a third of these patients, side effects are derived from cutaneous vasodilation and are primarily manifested by ankle edema. Calcium antagonists constitute one of the best options for a combined therapy. They have beneficial and synergistic effects when combined with RAS blockers (angiotensin-converting enzyme (ACE) inhibitors or angiotensin II receptor blockers (ARBs)) and in the case of dihydropyridines, with beta blockers. In both cases, an additional antihypertensive benefit is achieved. Furthermore, the metabolic side effects of diuretics are avoided.

5.3. Inhibitors of the converting enzyme of angiotensin

The mechanism of action consists of the inhibition of angiotensin II formation from angiotensin I. ACEIs also cause a decrease in aldosterone secretion induced by angiotensin II and inhibit the breakdown of bradykinin to increase levels of the mentioned vasodilator peptide. They are currently considered a first step for drug treatment in hypertension, and have demonstrated their ability to prevent cardiovascular events in hypertensive patients. Combined therapies with thiazide and loop diuretics are especially effective as they prevent the formation of angiotensin II-induced activation of renin secretion. One of its biggest advantages is that they are effective and safe in associated diseases, prolong survival, and reduce complications of patients with infarction and ventricular dysfunction. Likewise, these are required drugs in the treatment of heart failure and chronic renal failure with proteinuria. Finally, they are especially indicated in hypertension associated to diabetes mellitus due to its ability to prevent micro- and macrovascular coronary disease and nephropathy progression of secondary complications. Side effects are mainly the appearance in some patients of refractory therapeutically nonproductive dry cough. There have also been reported isolated cases of angioedema, which appears with the first doses, probably reflecting a phenomenon of hypersensitivity.

5.4. Antagonists angiotensin II receptor

These drugs reduce the inhibition of the renin-angiotensin system by the specific antagonism of angiotensin I and II. The advantage compared to ACE inhibitors is that it is better tolerated. A unique feature is a major effect on stroke prevention.

6. Hypertension in different biological conditions

Treatment of hypertension in special situations deserves special consideration because of its high prevalence and its added cardiovascular risk [3–6]:

6.1. Elderly

In hypertensive patients over 80 years, treatment should start from equal or higher levels of 160/90 and with the goal of achieving values of 150/90. From this recommendation are excluded hypertensive patients with great fragility, high levels of dependence, or debilitating diseases in which treatment decisions should be individualized.

6.2. Diabetes mellitus

It is a situation of high cardiovascular risk. In addition, diabetic nephropathy is a serious health problem that can be the leading cause for a need of dialysis and kidney transplant. Prognosis of diabetic patients depends on many factors besides glycemic control, among which figures blood pressure. The basic objectives in this case are the reduction of blood pressure to levels below 140/90, and strict metabolic control of the glycosylated hemoglobin to levels lower than 7%, in addition to low-density lipoprotein (LDL) cholesterol levels below 100 mg. The drug treatment for diabetic patients should be based on a serotonin-releasing agent (SRA) blocking agent, especially because of its renal protective effect, although with most patients it will be necessary to use two or more associated drugs. In this case, the groups to combine would be calcium-channel blockers and diuretics.

6.3. Ischemic heart disease

The objective of the control in these cases is to achieve figures lower than 140/90. Regardless of age, beta-blockers are the drugs of choice in patients with hypertension and angina. Calcium antagonists are an alternative in case of contraindication or intolerance to beta-blockers. It has also been found that spironolactone associated with beta-blockers reduces mortality in heart failure. There are studies supporting the indication of ACE inhibitors or ARBs in all patients with ischemic heart disease.

6.4. Cerebrovascular disease

Primary prevention constitutes one of the main expressions of the effectiveness of antihypertensive treatment control. Objectives are set here at values below 140/90 mmHg. Stroke especially affects the elderly hypertensive and depends markedly on the figures of control of the blood pressure. In clinical studies, it has been observed that diuretics and calcium antagonists have a special preventive capacity.

6.5. Kidney disease

Hypertensive patients with albuminuria kidney disease or reduction of renal filtrate below 30 ml/min should be considered at high risk for injury in the target organ.

6.6. Resistant hypertension

It is defined as the inability to achieve the desirable blood pressure despite modifications in lifestyle, the use of a diuretic, and at least two other blood pressure medicines. The prevalence is between 10 and 15%. Diagnosis requires confirmation by ambulatory monitoring (MAPA), since even a third of these patients present a white-coat phenomenon. Many patients will need the administration of a fourth type of drug. A small group of patients maintains a high blood pressure despite complex therapeutic regimens often with up to six or seven drugs and by ruling out other causes responsible for resistance to treatment. Recently, some nonpharmacological therapeutic techniques have been developed that have been proved effective in reducing blood pressure in these uncontrolled patients, such as renal sympathetic denervation, stimulation of carotid baroreceptors, or implantation of an arteriovenous fistula.

7. Exercise for hypertension

Guidelines for the management of hypertension recommend lifestyle modifications, in particular, a combination of diet and exercise. Nonpharmacological strategies are aimed to change the lifestyle and should be instituted in hypertensive and pre-hypertensive patients. Supervised exercise, the DASH diet, and weight reduction, combined with pharmacological therapy, are effective treatments for normalization of blood pressure. This statement is based on the knowledge that exercise is a landmark in primary prevention and coadjuvant in the treatment of hypertension [7]. Exercise is considered as a polypill and, thus, is necessary to know the optimal dosage (posology) in order to obtain maximal benefits [8]. Specifically, the response of blood pressure to exercise aims to ensure adequate blood flow to active muscles. BP depends on cardiac output (Q), which varies depending on myocardial contractility, heart rate (HR), blood volume, and peripheral vascular resistance. During the development of training, changes occur at the cardiovascular level, including increased systolic blood pressure, diastolic blood pressure, and mean arterial pressure (MAP) [9, 10].

The possible mechanisms of the antihypertensive effects of exercise are not fully clear but may include [7] (a) reduction in sympathetical-induced vasoconstriction and reduced catecholamine segregation; (b) increased insulin sensitivity; (c) anti-inflammatory effects; and (d) vascular structural adaptations.

Different international organizations such as the American Heart Association, American College of Cardiology, American College of Sports Medicine, European Society of Hypertension, European Society of Cardiology, or the Canadian Hypertension Education Program have developed their exercise prescription guidelines for adults with hypertension (for a review, see the original statements or Pescatello et al. [11]). Because the ideal dose is still unknown, the emergence of new modalities of exercise, in this chapter updated exercise strategies for managing and preventing hypertension, is presented. The modalities of exercise that have shown a relatively stronger supporting evidence are aerobic exercise (Class I, level of evidence A), dynamic resistance exercise (Class IIA, level of evidence B), and isometric handgrip (Class IIB, level of evidence C) [3].

7.1. Cardiovascular exercise

Low aerobic fitness has been proved to be a strong predictor for future cardiovascular disease and all-cause mortality in people with hypertension [12]. Not surprisingly therefore, there still seems to be a consensus that aerobic exercise should be prescribed as the primary type of exercise for the prevention, treatment, and control of hypertension [11]. Aerobic training can be performed either at continuous intensity or by short intervals of high-intensity exercise.

7.1.1. Continuous aerobic

The most extensive evidence with endurance training for hypertension is available for the continuous modality, understood as exercise of a constant sustainable intensity that is carried out over a given period of time, usually long.

7.1.1.1. Benefits of continuous aerobic training for hypertension

An acute session of continuous aerobic exercise can lower BP during the post-exercise period, known as the post-exercise hypotension (PEH). PEH is considered a prolonged decrease in resting blood pressure in the minutes and hours (up to 24 h) after an acute bout of exercise [13]. This decrease is more pronounced in hypertensive patients [14] and has also been shown in very old adults [15].

Chronic exercise results in more sustained reductions in BP, known as the exercise-training response [16]. According to different reviews and meta-analysis, aerobic exercise reduces BP in 5–7 mmHg [17]. However, it should also be noted that there is a considerable heterogeneity of the individual response to exercise training, attributed to environmental and genetic factors. About 20–25% of hypertensive patients could be nonrespondents to exercise [18].

7.1.1.2. *Mechanisms of benefits*

PEH is mediated by both central and peripheral factors. The fall in blood pressure seen after exercise seems to be caused by a reduced signal transduction from sympathetic nerve activation into vasoconstriction, and also local vasodilatator mechanisms. Recent evidence suggests that muscle afferents may play a major role in this response [19].

Chronic aerobic exercise in hypertensive patients may lead to a reduction in left ventricular mass [16], although results are not always conclusive. Decreased oxidative stress and inflammation levels, as well as an improved autonomic function, have been described as key mechanisms that contribute to the physiological benefits obtained with exercise. However, further research is needed to elucidate the exact mechanisms involved in the adaptations to chronic aerobic exercise [16].

7.1.1.3. *Dose of exercise*

When it comes to prescribe physical activity, the dose-response question is a very relevant issue. It refers to the relationship between different doses of physical activity and changes in a defined health parameter. The total physiological load is calculated for continuous endurance training as the product of the intensity, duration, and frequency of the exercise. There is still an ongoing debate about the optimal training dose, frequency, duration, and intensity of exercise for hypertensive patients.

In regard to intensity, few studies have actually compared the effects of different intensities of endurance exercise in blood pressure. Cornelissen et al. [20], using a randomized crossover design, compared the effects of two different programs of endurance training in BP, HR, and heart rate variability (HRV) in a group of 36 healthy sedentary older adults (mean age: 59 years, range: 55–71 years) with SBP of ≥120 mmHg or DBP of ≥80 mmHg. The study was composed of three 10-week periods. In the first and third period, the participants exercised at 33 or 66% of HR reserve (3 days per week, 50 min per session) in random order, with a sedentary period in between. The results of this study showed that endurance training at both lower and higher intensities reduced SBP significantly ($P < 0.05$) at rest, before exercise, during exercise, and during recovery to a similar extent. The effect of training on HR was more pronounced ($P < 0.05$) with higher intensity, whereas HRV was unaffected by the intervention.

However, more recent evidence suggests that moderate to high intensities produce higher BP reductions than low-intensity endurance training (<40% of heart rate reserve or <55% of maximum heart rate) [17].

On the other hand, the most appropriate duration and frequency of the exercise programs is not fully clear either, but most evidence is in line with the American College on Sports Medicine's recommendations of performing aerobic exercise most, preferably all, days of the week. The duration of each exercise session should range from 30 to 60 min [11].

In parallel, Cornelissen and Smart [17] concluded, after conducting a systematic review and meta-analysis, that exercise programs longer than 210 min per week produced the smallest reductions in BP compared to shorter programs, but probably due to the lower intensity of longer programs.

7.1.2. *High-intensity interval training*

There is increasing evidence showing that high-intensity training (85–100% of peak oxygen uptake [VO_{2peak}]) in hypertensive patients may elicit higher CVD benefits, including lowering blood pressure, compared to low- and moderate-intensity training [20, 21]. High-intensity interval training (HITT) is the most common form of high-intensity training, and consists of short bouts of high-intensity exercise interspersed with resting periods of similar, lower, or longer duration.

7.1.2.1. *Benefits of HIIT for hypertension*

Repeated bouts of high-intensity interval exercise (>85 VO_{2max}) have been shown to induce similar PEH responses to those obtained with continuous modalities of endurance training performed at lower intensities (~60 to 65% VO_{2max}), both when matched for exercise volume [22], and when compared to lower volume but performed at maximal intensity [23].

A 12-week HIIT program (85–90% of VO_{2max}) lowered blood pressure, improved cardiac function and aerobic capacity, and reduced the mean HR, with superior effects than continuous aerobic training performed at a lower intensity (60% of VO_{2max}) [21].

7.1.2.2. *Mechanisms of benefits*

The mechanisms of the benefits elicited by HIIT programs are probably similar to the ones described in the continuous modality. However, higher intensities elicit exponentially greater increases in sympathetic activity, so it is possible that it may produce higher autonomic adaptations. It has also been suggested that the larger shear stress provided by high-intensity exercise compared to moderate intensity programs further enhances the endothelial function, which plays a major role in the regulation of vasomotor tone and the development of atherosclerosis, and it has also been shown to be a strong predictor of CV risk in hypertensive patients [21–25].

7.1.2.3. *Dose of exercisee*

When it comes to HIIT training, the most important variables to consider are the duration of the short bouts of high-intensity exercise as well as the duration of the resting periods. More research is needed for establishing the optimal ratio of exercise and rest in people with hypertension, with protocols ranging from 4 × 4-min interval at 85–90% of VO_{2max} with a 3-min active pause between bouts [21] to two sets of 10 min composed of repeated phases of 15 s at 100% interspersed by 15 s of passive recovery, with another 4 min of passive recovery between sets [25].

In patients with coronary artery disease and heart failure, it has been suggested that short bouts of high-intensity exercise (100% of maximal aerobic power) such as 15 s for people with CAD and 30 s for people with HF (with passive recovery phases of the same duration between bouts, ratio 1:1) are optimal for obtaining benefits and improving adherence [26]. It is possible that these combinations are also adequate for people with hypertension as shown by Sosner et al. [25], but more research is needed in order to establish the most appropriate training regime.

7.2. Resistance exercise: dynamic resistance training

7.2.1. Benefits of dynamic resistance training for hypertension

Resistance exercise is increasingly recommended because studies have shown a reduced systolic blood pressure of 1.8 and 3.3 mmHg in diastolic blood pressure with this type of training [17].

7.2.2. Mechanisms of benefits

Initially, SBP and DBP rise because of the exercise pressor reflex to the cardiovascular center in the medulla from proprioceptors (mechanoreceptors and metaboreceptors) in active muscles. Arterial pressure rises to overcome the resistance to muscle perfusion caused by elevated intramuscular pressure interrupting arterial blood flow.

A hypotensive response has been reported several hours after exercise, caused by reduced norepinephrine levels and thus by the inhibition of sympathetic activity, a reduction in circulating angiotensin II, adenosine, and endothelin levels and their receptors in the central nervous system leading to decreased pulmonary vascular resistance (PVR), and increased baroreflex sensitivity. Hypotension is also triggered by the vasodilator effect of prostaglandins and nitric oxide.

7.2.3. Dose of exercise

As a general recommendation, Pescatello et al. [11] recommend to carry out a strength-training program of moderate intensity (i.e., 60–80% of 1 RM) consisting of eight to 10 exercises targeting the major muscle groups (performing multisets of 10–12 reps), and using different means of resistance training. This training should ideally be repeated 2–3 days per week.

Usually, it has been recommended to perform strength training dynamically, perhaps to avoid an increase in blood pressure. This effect is possibly due to a mechanical compression, causing increased peripheral resistance. Such resistance is related to blood viscosity and the length and radius of the vessel during muscle contraction in exercised muscle groups. This compression phenomenon that obstructs blood flow is associated with a reflex vasoconstriction in unexercised body areas that elevate BP. Moreover, when holding the breath while performing strength training (usually when the load is very high), an increase in the intrathoracic pressure caused by what is known as the Valsalva maneuver (MV) occurs [27]. Fleck et al. [28] observed that the SBP, DBP, and the double product (DP) increase during the last repetitions (one to three) of the series leading to volitional and muscular failure, regardless of the intensity. It appears that both, the prolongation of effort and the degree of reduction in execution speed (increased muscle static component) as the last few repetitions approach, cause an increase in BP regardless of the intensity, especially when performing sets to failure. In the same way, Wilborn et al. [10] demonstrated that the BP is significantly higher after 85% of 1 RM than 65% of that 1 RM. They found no significant differences in the response of SBP between exercising at 65 and 85% of 1 RM. The results of this study suggest that reasons other than mechanical compression and Valsalva maneuver may be the cause of the increase in BP due to the fact that 1 RM caused minor responses in BP. The study concluded that the longer duration of exercise at 65% of 1 RM is most likely to cause the largest increase in SBP and HR at this intensity.

7.3. Resistance exercise: isometric resistance training

Over the past 5 years, emerging evidence supports the potential use of isometric training for hypertension. Nevertheless, the meta-analysis of Cornelissen and Smart [17] highlights that isometric exercise shows the greater reductions in systolic and diastolic blood pressure, 10.9 and 6.2 mmHg, respectively, compared to other types of training. In a recent meta-analysis, Inder et al. [29], using a meta-analysis, have shown that isometric exercise will be more efficient in reduction on hypertension in males of >45 years old, using arms exercises unilaterally over 8 weeks of training.

In regard to the optimal dose for this type of training, Millar et al. [30] have reported that effective interventions should be characterized by:

- 4 × 2 min (1–4 min rest) of 20–50% of the maximal voluntary contraction.
- 3–4 days/week during at least 4–10 weeks.
- Being able to be carried out by lower limb exercises (i.e., leg press or squat) or upper limb (i.e., handgrip).

Finally, it should be noted that there is emerging evidence that shows significant improvements in blood pressure after physical-training programs with vascular occlusion superimposed, although the mechanisms of action that could support these improvements are currently unknown and therefore it is very difficult to establish appropriate recommendations for this type of exercise [31].

Some aspects that should be considered in connection with the exercise in people affected by hypertension are as follows:

- Progression should be gradual, avoiding large increases, especially in the intensity variable.
- Do not begin an exercise program if the person presents values of arterial pressure of 180/105 mmHg or higher.
- Monitor blood pressure during (or at the end of a set) the exercise by the average of two consecutive measurements spaced by 30 s.
- Use a cooldown phase of at least 5–10 min, in order to avoid an excessive post-exercise hypotension effect.

http://dx.doi.org/10.5772/65035

- If the response to exercise is hypotensive, that is, an inability to increase blood pressure despite increased demands for physical exercise, usually less than 20–30 mmHg, exercise must stop. This phenomenon may reflect the need for adjustment of the antihypertensive treatment.
- Monitor the pressor effect of an exercise at a given intensity. This factor is calculated by multiplying the systolic blood pressure for heart rate registered after exercise to a specific intensity.
- Provide a sufficient recovery time, at least 30 s, to return to baseline conditions of the cardiovascular system.
- Avoid exercise in which the head level is below the hips (declined exercises).
- Avoid high number of repetitions and muscular failure in resistance training.

8. Case study

Age: 55 years **Sex:** Male

BP: 155/110 mmHg. **Medication:** Treated with diuretics for 3 months.

Physical activity levels: Sedentary.

Body mass index (BMI): 33 kg/m^2.

Patient history: No familiar antecedents of cardiovascular accident.

The proposed exercise would be summarized as follows:

		Aerobic exercise	Isometric exercise
Initial stage	Fisrt week	**Walking** At least 3 days/week Accumulated 3 set of 10 minutes Low-moderate intensity (40-60 % HR reserve)	**Handgrip exercise** 2 days/week 2 sets x 1 minute (20-30% 1RM)
	Second week	**Walking** At least 5 days/week Acumulated 3 set of 10 minutes Low-moderate intensity (40-60 % HR reserve)	**Handgrip exercise** 2 days/week 2 sets x 2 minute (20-30% 1RM) **Squat** 2 days/week 2 sets x 1 minute (20-30% 1RM)

		Aerobic exercise	Isometric exercise	Dynamic resistance training
Progression stage	Third week	**Walking or cycling** 2 days / week 30 minutes Low-moderate intensity (40-60 % HR reserve) 3 days/week Accumulated 3 set of 10 minutes Low-moderate intensity (40-60 % HR reserve)	**Handgrip exercise** 3 days/week 2 sets x 1 minute (20-30% 1RM) **Squat** 2 days/week 3 sets x 1 minute (20-30% 1RM)	**Core exercises** 1.Superman 2.Lateral bridge 3.Modified curl-up 2-3 sets x 10-20 seconds **Upper limb exercises** 4.Chest Press 5.Row machine **Lower limb exercises** 6.Hamstrings machine 7.Calves machine 8.Abductor machine 2-3 sets x 10 – 12 reps (15) (40-50% 1RM)
	Fith week	**Walking or cycling** 2 days / week 30 minutes Low-moderate intensity (40-60 % HR reserve) 2 days / week 40 minutes Low-moderate intensity (40-60 % HR reserve) 1 days/week Accumulated 3 set of 10 minutes Moderate-high intensity (60-80 % HR reserve)	**Handgrip exercise** 3 days/week 2 sets x 2 minute (20-30% 1RM) **Squat** 2 days/week 4 sets x 1 minute (20-30% 1RM)	**Core exercises** 1.Superman 2Lateral bridge 3.Modified curl-up 2-3 sets x 20-30 seconds **Upper limb exercises** 4.Chest Press 5.Row machine **Lower limb exercises** 6.Leg press 7.Hamstrings machine 8.Calves machine 9.Abductor machine 2-3 sets x 10 – 12 reps (15) (50-60% 1RM)

http://dx.doi.org/10.5772/65035

		Aerobic exercise	Isometric exercise	Dynamic resistance training
Progression stage	Sixth week	**Walking or cycling** 2 days / week 40 minutes Low-moderate intensity (40-60 % HR reserve) 2 days / week 50 minutes Low-moderate intensity (40-60 % HR reserve) 1 days/week Acumulated 3 set of 10 minutes Moderate-high intensity (60-80 % HR reserve)	**Handgrip exercise** 4 days/week 3 sets x 1 minute (20-30% 1RM) **Squat** 2 days/week 3 sets x 2 minute (20-30% 1RM)	**Core exercises** 1.Superman 2Lateral bridge 3.Modified curl-up 2-3 sets x 30-40 seconds **Upper limb exercises** 4.Chest Press 5.Row machine 6.Shoulder press **Lower limb exercises** 7.Lunge 8.Hamstrings machine 9.Calves machine 10.Abductor machine 2-3 sets x 10 – 12 reps (15) (60-70% 1RM)
	Tenth week	**Walking or cycling or jogging or swimming** 2 days / week 50 minutes Low-moderate intensity (40-60 % HR reserve) 2 days / week 60 minutes Low-moderate intensity (40-60 % HR reserve) 1 days/week 10-30 minutes High Intensity Exercise High intensity (>80 % HR reserve)	**Handgrip exercise** 4 days/week 4 sets x 1 minute (20-30% 1RM) **Squat** 2 days/week 4 sets x 2 minute (20-30% 1RM)	**Core exercises** 1.Superman 2.Lateral bridge 3.Modified curl-up 2-3 sets x 40-50 seconds **Upper limb exercises** 4.Chest Press 5.Row machine 6.Shoulder press 7.Triceps machine **Lower limb exercises** 8.Lunge 9.Hamstrings machine 10.Calves machine 2-3 sets x 10 – 12 reps (15) (60-80% 1RM)

		Aerobic exercise	Isometric exercise	Dynamic resistance training
Maintain stage	>Eleventh week	**Walking or cycling or jogging or swimming** At least 4 days / week 60 minutes Low-moderate intensity (40-60 % HR reserve) 1 days/week Acumulated 3 set of 10 minutes Moderate-high intensity (60-80 % HR reserve)	**Handgrip exercise** 4 days/week 4 sets x 2 minute (20-30% 1RM) **Squat** 2 days/week 4 sets x 2 minute (20-50% 1RM)	**Core exercises** 1.Superman 2.Lateral bridge 3.Modified curl-up 2-3 sets x 50-60 seconds **Upper limb exercises** 4.Chest Press 5.Row machine 6.Shoulder press 7.Triceps machine 8.Biceps machine **Lower limb exercises** 9.Lunge 10.Deadlift (without Valsalva maneuver) 2-3 sets x 10 – 12 reps (15) (60-80% 1RM)

Author details

Chulvi-Medrano Iván[1*], Sanchis-Cervera José[2], Tortosa-Martínez Juan[1] and Cortell-Tormo Juan Manuel[1]

*Address all correspondence to: ivan.chulvi@ua.es

1 Department of General and Specific Didactics, University of Alicante, Alicante, Spain

2 Department of General and Specific Didactics, Emeritus Hematology Hospital de la Plana, Castellón, Spain

References

[1] Kearney PM, Whelton M, Reynolds K, et al. Global burden of hypertension: analysis of worldwide data. Lancet. 2005;365:217–223.

[2] Llisterri JL, Alonso FJ, Gorostidi M, et al. Differences between office and ambulatory control of hypertension in very elderly patients. The CARDIORISC-MAPAPRES project. Med Clin (Barc). 2009; 133(20): 769–776.

[3] Brook RD, Appel LJ, Rubenfire M, et al. Beyond medications and diet: alternative approaches to lowering blood pressure. A scientific statement from the American Heart Association. Hypertension. 2013;61:1360–1383.

[4] Weber MA, Schiffrin EL, White WB, et al. Clinical practice guidelines for the management of hypertension in the community a statement by the American Society of Hypertension and the International Society of Hypertension. J Hypertens. 2014;32:3–15.

[5] James PA, Oparil S, Carter BL, et al. 2014 evidence-based guidelines for the management of high blood pressure in adults: report from the panel members appointed to the Eighth National Committee. JAMA. 2014; 3011:507–520.

[6] Khan NA, McAlister FA, Lewanczuk RZ, et al. The 2005 Canadian Hypertension Education Program recommendations for the management of hypertension therapy. Can J Cardiol. 2005;21:657–672.

[7] Pedersen BK, Saltin B. Exercise as medicine—evidence for prescribing exercise as therapy in 26 different chronic disease. Scand J Med Sci Sports. 2015;25(Suppl 3):1–72.

[8] Pareja-Galeano H, Garatachea N, Lucia A. Exercise as a polypill for chronic diseases. Prog Mol Biol Transl Sci. 2015;135;497–526.

[9] MacDougall JD, Tuxen D, Sale DG, et al. Arterial blood pressure response to heavy resistance exercise. J Appl Physiol. 1985;58(3):785–790.

[10] Wilborn C, Greenwood M, Wyatt F, et al. The effects of exercise intensity and body position on cardiovascular variables during resistance exercise. JEPonline. 2004;7(4): 29–36.

[11] Pescatello LS, MacDonald HV, Lamberti L, et al. Exercise for hypertension: a prescription update integrating existing recommendation with emerging research. Curr Hypertens Rep. 2015;17:87.

[12] Kokkinos P, Manolis A, Pittaras A, Doumas, et al. Exercise capacity and mortality in hypertensive men with and without additional risk factors. Hypertension. 2009;53(3): 494–499.

[13] MacDonald JR. Potential causes, mechanisms, and implications of post exercise hypotension. J Hum Hypertens. 2002;16:225–236.

[14] Cardoso Jr CG, Gomides RS, Queiroz AC, et al. Acute and chronic effects of aerobic and resistance exercise on ambulatory blood pressure. Clinics (Sao Paulo). 2010;65:317–325.

[15] Oliveira J, Mesquita-Bastos J, Argel de Melo C. Postaerobic exercise blood pressure reduction in very old persons with hypertension. J Geriatr Phys Ther. 2016;39(1):8–13.

[16] Hegde SM, Solomon SD. Influence of physical activity on hypertension and cardiac structure and function. Curr Hypertens Rep. 2015;17:77.

[17] Cornelissen VA, Smart NA. Exercise training for blood pressure: A systematic review and meta-analysis. J Am Heart Assoc. 2013;2(1): 1–9. doi: 10.1161/JAHA.112.004473.

[18] Pescatello LS, Blanchard BE, Tsongalis GJ, et al. A comparison of the genetic and clinical profile of men that respond and do not respond to the immediate antihypertensive effects of aerobic exercise. Appl Clin Genet. 2008;1:7–17.

[19] Halliwill JR, Buck TM, Lacewell AN. Postexercise hypotension and sustained postexercise vasodilatation: what happens after we exercise? Exp Physiol. 2013;98:7–18.

[20] Cornelissen VA, Verheyden B, Aubert AE, et al. Effects of aerobic training intensity on resting, exercise and post-exercise blood pressure, heart rate and heart-rate variability. J Hum Hypertens. 2010;24:175–182.

[21] Molmen-Hansen HE, Stolen T, Tjonna AE, Aamot, et al. Aerobic interval training reduces blood pressure and improves myocardial function in hypertensive patients. Eur J Prev Cardiol. 2012;19:151–160.

[22] Lacombe SP, Goodman JM, Spragg CM, et al. Interval and continuous exercise elicit equivalent postexercise hypotension in prehypertensive men, despite differences in regulation. Appl Physiol Nutr Metab. 2011;36:881–891.

[23] Rossow L, Yan H, Fahs CA, et al. Postexercise hypotension in an endurance-trained population of men and women following high-intensity interval and steady-state cycling. Am J Hypertens. 2010;23:358–367.

[24] Swain DP, Franklin BA. Comparison of cardioprotective benefits of vigorous versus moderate intensity aerobic exercise. Am J Cardiol. 2006;97(1):141–147.

[25] Sosner P, Gayda M, Dupuy O, et al. Ambulatory blood pressure reduction following high-intensity interval exercise performed in water or dryland condition. J Am Soc Hypertens. 2016; DOI: 10.1016/j.jash.2016.02.011.

[26] Guiraud T, Juneau M, Nigam A, et al. Optimization of high intensity interval exercise in coronary heart disease. Eur J Appl Physiol. 2010;108:733–740.

[27] MacDougall JD, McKelvie RS, Moroz DE, et al. Factors affecting blood pressure during heavy weight lifting and static contractions. J Appl Physiol. 1992;73(4):1590–1597.

[28] Fleck SJ, Dean LS. Resistance-training experience and the pressor response during resistance exercise. J Appl Physiol. 1987;63(1):116–120.

[29] Inder JD, Carlson DJ, Dieberg G, et al. Isometric exercise training for blood pressure management: a systematic review and meta-analysis to optimize benefit. Hypertens Res. 2016;39:88–94.

[30] Millar PJ, McGowan CL, Cornelissen VA, et al. Evidence for the role of isometric training on reducing blood pressure: potential mechanisms and future directions. Sports Med. 2013; DOI: 10.1007/s40279-013-0118-x.

[31] Chulvi-Medrano I. Resistance training with blood flow restriction and hypertensive subjects. J Hum Kinet. 2015;27(46):7–8.

Chapter 6

Effects of Resistance Training on Autonomic Nervous Function in Older Individuals

Hidetaka Hamasaki

Additional information is available at the end of the chapter

http://dx.doi.org/10.5772/65389

Abstract

Skeletal muscle mass, strength, and function decline with aging are the symptoms that characterize sarcopenia, which has become a significant problem in aging societies. Aging is also associated with arterial stiffness and autonomic nervous dysfunction, leading to increase in the risk of cardiovascular disease (CVD) and mortality. Resistance training (RT) is effective for improving muscle fitness in older individuals as well as young healthy individuals. However, the effects of RT on autonomic nervous function (ANF) in the elderly have not been fully elucidated. The author reviewed the current evidence regarding RT and ANF in older individuals. Whole-body, high-intensity or progressive RT had either no effect on ANF or perhaps an unfavorable effect on ANF. On the other hand, local isometric, moderate-intensity RT may have a beneficial effect on ANF in older individuals. The combination of RT and aerobic exercise had a favorable effect on ANF in older patients with comorbidities. However, the optimal intensity, frequency, and duration of RT for improving ANF in older individuals remain unknown. Further studies with a large number of subjects are warranted.

Keywords: resistance training, sarcopenia, autonomic nervous function, heart rate variability

1. Introduction

Sarcopenia, age-related loss of muscle mass and/or decline in muscle strength and performance [1] has become a significant problem in aging societies. Sarcopenia is associated with the increased risk of disability [2] and mortality [3], and resistance training (RT) should be performed in the elderly as well as healthy young individuals. A cohort study showed that guideline-concordant RT reduced the risk of all-cause mortality (odds ratio 0.64) in older individuals over 65 years of age [4]. On the other hand, we should be careful when prescribing RT for older individuals with disabilities or diseases because such individuals may develop physical impairments due to their low physical fitness. Adverse effects of RT on the elderly have not been adequately reported in previous clinical studies [5]. Transient changes in blood pressure in response to exercise or environmental changes (e.g., posture, atmospheric pressure, temperature) are intricately regulated by the reciprocal action of the sympathetic and parasympathetic nervous systems. However, the incidences of orthostatic and postprandial hypotension increase with age [6], and systolic hypertension is caused by an increase in arterial stiffness and sympathetic nervous dysfunction in the elderly [7]. Aging is also a significant factor associated with reduced baroreflex sensitivity and increased blood pressure [7], and furthermore, baroreflex dysfunction in older subjects may be the underlying pathophysiological mechanism in vasovagal syncope [8]. Muscle sympathetic nerve activity (MSNA) increases with age in both men and women [9]. Higher sympathetic nerve activity causes hypertension with aging. Long-term endurance exercise (EE) decreases sympathetic nerve activity and increases parasympathetic nerve activity, which may reduce the risk of cardiovascular disease (CVD) [10]. However, the effect of RT on the cardiac autonomic nervous function (ANF) of younger individuals is controversial. Progressive RT with an emphasis on increasing the muscle strength of the lower limbs improved the vagal modulation of heart rate [11], whereas whole-body RT had no effect on vagal-cardiac control or cardiovagal baroreflex sensitivity [12] and heart rate variability (HRV) [13]. On the other hand, Taylor et al. [14] showed that isometric handgrip training at a moderate intensity could reduce resting blood pressure and increase vagal modulations in older individuals with hypertension. A previous review also showed that isometric exercise training reduces systolic and diastolic blood pressure in both young and older individuals [15]; however, the effects of RT on autonomic nervous function (ANF), especially in older individuals, are not fully elucidated. Thus, the aim of this review was to summarize and evaluate the current evidence of RT and ANF and to provide clinicians with knowledge on the role of RT in ANF among middle-aged and older individuals.

2. Methods

The author searched for English literature on RT and ANF by using the PubMed and MEDLINE databases. The search terms were "resistance training or resistance exercise," "autonomic nervous function," and "aged or elderly." The search returned 54 published articles. Studies were included if they met the following criteria: (1) it was a randomized controlled trial, (2) the participants' mean age ≥55 years, and (3) the study duration ≥12 weeks. Studies of experimental animals were excluded from this review. The titles and abstracts of the identified articles were reviewed to determine their relevance. A total of eight articles were eligible.

3. Heart rate variability (HRV)

HRV is the most common noninvasive analysis of cardiac autonomic nervous function. Time-domain measurements of HRV include the standard deviation of normal intervals, the root-mean-square of successive differences, and the proportion of R-R intervals that differ by >50 ms from the previous R-R interval [16]. A decrease in these measurements represents the impairment of ANF in CVD [17]. The total power of HRV is an indicator of autonomic nervous system activity. The low-frequency (0.04–0.15 Hz) component of HRV is mediated by sympathetic and parasympathetic nerve activities, and the high-frequency (0.15–0.40 Hz) component of HRV is a marker of cardiac parasympathetic nervous function. The ratio of low frequency to high frequency indicates the predominance of sympathetic nerve activity [16]. HRV decreases with age as a result of reduced parasympathetic nervous system activity and the predominance of sympathetic modulation [18]. Therefore, HRV is useful for assessing ANF in older individuals.

4. The effects of RT on ANF in healthy older subjects

ANF (cardiovascular sympathetic and parasympathetic modulations) is usually evaluated by the spectral analysis of R-R intervals of electrocardiogram and blood pressure variabilities [19]. Resting HRV is also useful in the diagnosis of cardiac autonomic neuropathy in patients with type 2 diabetes [20]. Previous studies have revealed that HRV can predict mortality in healthy middle-aged and older individuals [21, 22] and patients with heart failure [23].

Madden et al. [24] examined the effects of endurance exercise (EE) and RT on HRV in healthy older women on the hypothesis that RT had no effect on HRV, while EE had a beneficial effect on HRV. Forty-five female subjects (mean age 69.9 ± 0.9 years) participated in the trial. Subjects with histories of angina, myocardial infarction, stroke, hypertension, diabetes, chronic pulmonary disease, orthopedic impairment, and medications were excluded. Subjects were randomly assigned to one of three groups: a RT group (n = 15), a EE group (n = 10), and a sedentary group (n = 15). The mean body mass index (BMI) of each group was 26.8 ± 1.5, 28.5 ± 2.2, and 28.2 ± 1.3 kg/m^2, respectively. The subjects engaged in each exercise intervention for 6 months. The RT program consisted of three sets with 8–12 repetitions of 10 exercises and was performed five times per week under the supervision of a certified trainer. The goal of the RT was to improve upper and lower muscle strength to 85% of a one-repetition maximum (1RM). On the other hand, the EE program consisted of moderate- to vigorous-intensity exercise on a cycle ergometer and was performed five times per week. The intensity of the EE was set to 50–60% of maximal heart rate for the first 2 months and progressed to 80–85% of maximal heart rate for the remaining 4 months. No significant change in VO_{2max} was found in the RT group, while VO_{2max} significantly increased in the EE group. After the 6-month intervention, the maximal heart rate did not change in both the RT and EE groups; however, the standard deviation of the 5-min mean R-R interval and the standard deviation of heart rate were significantly increased in the EE group. RT had no significant effect on measures of HRV. The authors speculated that the reason why RT did not affect HRV was that the increased arterial stiffness induced by RT might have reduced arterial baroreceptor sensitivity, resulting in a reduction in HRV. Indeed, high-intensity RT is known to be associated with worsening arterial stiffness [25]. In addition, increased arterial stiffness is inversely associated with cardiac vagal modulation [26]. High-intensity RT with breath holding may cause both sympathovagal imbalance (sympathetic nerve ascendant) and a decrease in arterial compliance in older subjects. The intensity of RT (progressing to 85% of 1RM) might have been too high to have a beneficial effect on ANF in the elderly.

Neuroendocrinological system	Changes
Growth hormone	↓
Insulin-like growth factor-I	↓
Thyroid stimulating hormone	↓
Testosterone	↓
Estrogen	↓
Glucocorticoids	↑
Corticotropin-releasing hormone	↑
Sympathetic nerve activity	↑

Table 1. Changes in neuroendocrine hormone concentrations in older individuals.

Melo et al. [27] showed that high eccentric RT reduced HRV. Twenty-eight healthy older men were recruited, but only nine subjects (mean age 62 ± 2.0 years) completed the study protocol. The mean BMI was 25.43 ± 1.92 kg/m^2. The subjects carried out 2–4 sets of bilateral eccentric knee flexion and extension 2 days per week for 12 weeks. The subjects performed 8–12 repetitions at an intensity of 75–80% of peak torque. After 12 weeks of training, systolic blood pressure decreased from 123.78 ± 8.3 mmHg to 117.67 ± 10.2 mmHg. The knee flexion and extension peak torque also increased by 13 and 20%, respectively. No changes in heart rate and log [root-mean-square of successive differences] were observed; however, the ratio of low frequency to high frequency significantly increased after training. In this study, there was a considerable discrepancy between the decrease in systolic blood pressure and the increase in HRV by 12-week RT. The authors described that the increase in HRV might have been caused by increased catecholamine levels [28, 29] and not by increased arterial stiffness [30]. However, it is unclear whether the change in catecholamine levels alone could explain the unfavorable effect on ANF. Although sympathetic nerve activity increases with age, tissue responsiveness to catecholamine decreases with age. Older individuals respond differently to exercise stress compared to younger individuals [31]. Growth hormone, insulin-like growth factor-I, thyroid stimulating hormone, testosterone, and estrogen production decline with advancing age [31–33]. In contrast, aging is associated with elevated basal levels of circulating glucocorticoids and central levels of corticotropin-releasing hormone, which are secreted in response to stress via activation of the hypothalamic pituitary adrenal axis [33]. Recent studies have suggested that leptin also induces the activation of the sympathetic nervous system [32]. Because various neuroendocrine hormones are altered in the elderly (**Table 1**) and they have mutual relationships with each other, it is difficult to conclude that increased catecholamine levels due to RT are responsible for the reduction in HRV.

Takahashi et al. [34] reinvestigated whether eccentric strength training affected HRV because their previous study [27] showed that eccentric strength training had an unfavorable effect on HRV. A total of 22 healthy elderly male subjects participated in the study. Subjects who had hypertension, diabetes, chronic obstructive pulmonary disease, cardiovascular, respiratory and neurological disease, and musculoskeletal disease were excluded. Three subjects in the training group and two subjects in the control group dropped out. The strength training program consisted of 2–4 sets of bilateral knee eccentric flexion and extension. The subjects in the training group performed RT with 8–12 repetitions per set at the intensity of 70–80% of the peak torque value 2 days per week for 12 weeks. The heart rate response and HRV were evaluated during submaximal isometric contraction of knee extension (15, 30, and 40% of the peak torque value). The mean ages of the subjects in the training group and the control group were 62 ± 2 and 64 ± 4 years, respectively. The mean BMIs of subjects in the training group and the control group were 25.5 ± 1.8 and 26.6 ± 2.6 kg/m^2, respectively. After 12 weeks of training, peak eccentric torque (extensor and flexor) significantly increased, and group effects were also observed. On the other hand, no significant time or group effects on heart rate change during submaximal isometric contraction were found; RT did not modify heart rate change. Moreover, no significant time or group effects were observed on HRV (the root-mean-square of successive differences index) during the submaximal isometric exercise. The heart rate response pattern and HRV in the training group were similar to those in the control group. Eccentric strength training did not modify HRV or the heart rate response during isometric exercise in older men. The authors described that HRV in older subjects may be less sensitive to RT because HRV decreases with advancing age [35]. However, the effects of eccentric RT on ANF cannot be affirmed. The difference between these two studies, Melo et al. and Takahashi et al., is the measurement method of heart rate and HRV. In the former study, HRV was measured at rest, whereas in the latter study, heart rate and HRV response were measured during isometric exercise. HRV during isometric exercise and after eccentric RT includes not only a chronic effect of eccentric RT but also an acute effect of isometric exercise. Isometric handgrip exercises at 30% of the maximal isometric voluntary contraction increase vagal modulation in older hypertensive subjects [14]. Taken together, isometric RT at a moderate intensity has a beneficial effect on ANF; however, eccentric RT at a high intensity has no effect or some unfavorable effects on ANF in older individuals.

Karavirta et al. [36] examined the effects of concurrent EE and RT on heart rate dynamics compared with EE or RT alone. They measured not only conventional HRV but also fractal heart rate dynamics, utilizing detrended fluctuation analysis. This method is considered to be more suitable for evaluating heart rate dynamics during exercise [37]. A total of 105 male subjects were randomly allocated to an EE group ($n = 23$), a RT group ($n = 25$), a combined EE and RT group ($n = 29$), and a control group ($n = 16$). Ninety-three subjects (mean age 55.6 ± 7.4 years) completed the 21-week intervention program. EE was performed twice a week, and all the sessions were supervised. The subjects trained for 30 min on a bicycle ergometer under the aerobic threshold during the first 7 weeks, and both the duration of each session and the percentage above the anaerobic threshold were increased during the weeks from 8 to 14 and from 15 to 21. RT was also carried out twice a week, and all the sessions were supervised. The RT program included 7–10 exercises per session. The subjects carried out RT with light loads (40–60% of 1RM) and a high number of repetitions (15–30) with three sets during the first 7 weeks. During weeks 8–14, the loads were increased up to 60–80% of 1RM with 6–12 repetitions per set. The subjects carried out RT with higher loads (70–85% off 1RM) and 5–8 repetitions per set to optimize the muscle strength during the last 7 weeks. The combined training group performed both twice weekly endurance and resistance trainings. After 21 weeks of training, VO_{2max} increased by 11.9 ± 11.0% in the EE group and 10.1 ± 9.8% in the combined training group, whereas no significant improvement in VO_{2max} was observed in the RT group. The 1RM of the bilateral leg extension significantly increased by 21.1 ± 7.9% in the RT group and by 22.1 ± 9.5% in the combined training group compared with the increase of 7.1 ± 5.1% in the EE group. Only subjects in the combined training group showed a decrease in fractal heart rate behavior during exercise, and the group effects were significant. The short-term fractal scaling exponent decreased from 1.18 ± 0.20 to 1.11 ± 0.21 in the combined training group during supine rest, whereas no significant change was observed in the EE or RT group. In addition, the relative change in the short-term fractal scaling exponent was inversely correlated with age at an exercise intensity of 90% of the maximal aerobic work rate. As for spectral measures of HRV, the decrease in resting heart rate was correlated with the changes in low-frequency power (0.04–0.15 Hz) and high-frequency power (0.15–0.40 Hz) in only the combined training group. This study showed that RT alone did not change heart rate dynamics. The ability to adapt to physiological changes is impaired in older individuals [38]. However, the improvement in adaptability to physiological stress, including exercise, is important because older individuals are at a high risk of CVD. Although no effects of RT on heart rate dynamics were found, the synergetic effect between EE and RT on fractal heart rate behavior was detected. The authors stated that the underlying mechanism was unknown; however, the change in MSNA due to exercise might be associated with the improvement in ANF. A previous study showed that MSNA response increased 1 week after RT using a handgrip exercise with maximal effort; however, the MSNA returned to the baseline level at 4 weeks post-training in healthy young individuals [39]. The enhanced MSNA response to strength exercise may be caused by the activation of the central nervous system rather than the muscle metaboreflex [39], indicating that sympathetic nerve hyperactivity is diluted by the repetition of RT. Long-term exercises, including RT, decreased MSNA and systolic arterial pressure and increased baroreflex sensitivity in patients with myocardial infarction [40], which suggests that the combination of EE and RT ameliorates blood vessel elasticity and arterial baroreflex control. MSNA may be a more useful and reliable indicator of ANF in the elderly than measures of HRV because it is representative of sympathetic nerve activity that innervates not only the heart but also the kidney and skeletal muscle, and the reproducibility of age and gender is very high [9].

Kanegusuku et al. [41] investigated the effects of high-intensity progressive RT on cardiovascular function and autonomic neural regulation in older individuals. Among the 83 subjects, 37 subjects were randomized into the RT group ($n = 18$) and the control group ($n = 19$). However, six subjects dropped out in each group, and a total of 25 subjects completed the study. Subjects in the RT group performed seven supervised exercises (horizontal leg press, bilateral knee flexion, plantar flexion on the horizontal leg press, unilateral hip extension, chest press, lat pull down, and upright row on the isoinertial machines) twice a week for 16 weeks. During the first 4 weeks, subjects carried out two sets of RT with 10 repetitions, and the training sets and repetitions were progressively increased every 4 weeks. The mean ages of subjects in the RT and control group were 63 ± 4 and 64 ± 4 years, respectively. The mean BMIs of subjects in the RT and control group were 26.8 ± 4.7 and 25.7 ± 4.2 kg/m^2, respectively. The quadriceps cross-sectional area and the strengths of the leg press and chest press significantly increased after the 16-week high-intensity, progressive RT compared with controls. However, cardiovascular and autonomic nervous variables, such as blood pressure, systemic vascular resistance, cardiac output, heart rate, the variance of R-R intervals, high frequency, low frequency, and spontaneous baroreflex sensitivity, did not have any significant effect or interaction, and no differences between groups existed. High-intensity, progressive RT could increase muscle mass and strength in healthy older individuals without a decrease in autonomic control. In older subjects, ANF might be less responsive to exercise stress [42] because their sympathetic nerve activity is already increased at rest [9]. Even if it is high intensity, whole-body RT may not worsen ANF, even though it cannot improve ANF in the elderly.

References	Subjects	RT	Results
Madden et al. [24]	40 female subjects Age: 69.9 ± 0.9 years Dropout: five subjects	10 exercises for both upper and lower body utilizing free weights 8–12 repetitions × 3 sets, 5 times per week for 6 months Progressed to 85% of 1RM	No effects on measures of HRV
Melo et al. [27]	Nine healthy male subjects Age: 62 ± 2.0 years Dropout or excluded: 19 subjects	Bilateral eccentric knee flexion and extension 8–12 repetitions with an intensity of 75–80% peak torque × 2–4 sets, twice a week for 12 weeks	No effects on resting heart rate and log RMSSD Low-frequency/high-frequency index↓ Systolic blood pressure↓
Takahashi et al. [34]	17 healthy male subjects Age: 62 ± 2 years in the training group and 64 ± 4 years in the control group Dropout: five subjects	Bilateral eccentric knee flexion and extension 8–12 repetitions with an intensity of 70–80% peak torque × 2–4 sets, twice a week for 12 weeks HRV was evaluated during submaximal isometric contraction of knee extension	No effects on heart rate response and RMSSD during submaximal isometric contraction of knee extension
Karavirta et al. [36]	93 untrained male subjects Age: 55.6 ± 7.4 years Dropout: 12 subjects	7–10 exercises for all the main muscle groups Twice a week for 21 weeks, First 7 weeks: 15–30 repetitions with 40–60% of 1RM Second 7 weeks: 6–12 repetitions	HRV↓ in the combined RT and EE group No effects on HRV in the RT alone group

References	Subjects	RT	Results
		with 60–80% of 1RM The last 7 weeks: 5–8 repetitions with 70–85% of 1RM	
Kanegusuku et al. [41]	25 subjects (seven men and 18 women) Age: 63 ± 4 years in the training group and 64 ± 4 years in the control group Dropout: 12 subjects	Seven exercises (horizontal leg press, bilateral knee flexion, plantar flexion on the horizontal leg press; unilateral hip extension, chest press, lat pull down, and upright row on the isoinertial machines) Twice a week for 16 weeks, 10 repetitions × 2 sets, overload was progressively increased; however, the exact intensity of each session was not described in detail(only training repetitions and sets were described)	No effects on blood pressure, systemic vascular resistance, cardiac output, heart rate, the variance of R-R intervals, high frequency, low frequency, and spontaneous baroreflex sensitivity
Wanderley et al. [43]	50 subjects (11 men and 39 women) Age: 67.3 ± 4.9 years in the RT group, 69.9 ± 5.7 years in the EE group, and 67.8 ± 5.5 years in the control group Dropout: 35 subjects	Nine exercises (leg press, chest press, leg extension, seated row, seated leg curl, abdominal flexion, biceps curl, low-back extension, and triceps extension) Three times per week for 8 months, 12–15 repetitions × 2 sets with an intensity of 50–60% of 1RM→resistance was progressively increased to 80% of the re-evaluated 1RM	Blood pressure↓ and resting heart rate↓ in the EE group No effects on HRV in both the EE and RT groups

RT, resistance training; ANF, autonomic nervous function; HRV, heart rate variability; 1RM, one-repetition maximum; RMSSD, root-means-square of successive differences; EE, endurance training.

Table 2. Clinical studies that investigated the effects of RT on ANF in older individuals.

Wanderley et al. [43] examined the effectiveness of EE and RT on body fat, ANF, and low-grade inflammation in community-dwelling older individuals. Eighty-five subjects were randomly allocated into an EE group, a RT group, or a control group. Subjects with acute illness, severe or uncontrolled hypertension, cardiovascular and/or respiratory disease, and pharmacological therapies that could affect cardiovascular function were excluded. Thirty-five subjects dropped out from the protocol, and a total of 50 subjects were analyzed: EE (n = 20), RT (n = 11), and control (n = 19). The subjects trained three times per week for 8 months. Each exercise session was performed for approximately 50 min under supervision. The subjects in the EE group performed a walking aerobic exercise in addition to stepping and dancing with 10 min of warm-up and cool-down. The intensity of exercise was 50–60% of heart rate reserve during the first month and increased up to 70–80% of heart rate reserve. The RT consisted of nine exercises (leg press, chest press, leg extension, seated row, seated leg curl, abdominal flexion, biceps curl, low-back extension, and triceps extension) and was performed with 12–15 repetitions per set. The subjects carried out two sets of 12–15 repetitions at 50–60% of 1RM for the first month, and resistance increased to 80% of the reevaluated 1RM every 2 months. The mean age of subjects was 67.3 ± 4.9 (RT), 69.9 ± 5.7 (EE), and 67.8 ± 5.5 (control) years. The mean BMI of subjects was 29.5 ± 5.0 (RT), 28.1 ± 4.1 (EE), and 27.2 ± 3.5 (control) kg/m^2. Fifty-six percent of subjects had hypertension, 42% had dyslipidemia, and 18% had diabetes. Body fat significantly decreased in the EE and RT group after the 8-month training. The EE group demonstrated significant decrease in systolic and diastolic blood pressure and resting heart rate. RT did not affect blood pressure, heart rate, lipid profile, inflammation markers, such as hs-CRP, TNF-α, and IL-6, and 6-min walk distance. No significant change in HRV was observed in all the groups. Whole-body RT had no effects on ANF. This study is important because the study participants were not limited to healthy older individuals. The study population is more similar to that of the real world compared with previous clinical studies. The number of chronic diseases, such as hypertension, diabetes, stroke, heart disease, and cancer, increases with age. Although the prevalences of chronic diseases vary substantially by country, more than 15% of older individuals have six or more diseases in the US [44]. Therefore, clinicians should consider the influence of comorbidities in the elderly patients on ANF. This review also focuses on clinical studies that investigated the effects of RT on ANF in older subjects with heart disease. **Table 2** shows a list of published articles that investigated the effects of RT on ANF in healthy older subjects.

5. The effects of RT on ANF in older subjects with heart failure and diabetes

Few studies have investigated the effects of RT alone on ANF in older subjects with comorbidities. Although they did not evaluate the isolated effects of RT, two studies assessed the effects of cardiac rehabilitation on autonomic control in patients with heart failure [45] and the effects of combined aerobic/resistance training in patients with type 2 diabetes [46]. Ricca-Mallada et al. [45] evaluated the effects of cardiac rehabilitation on deceleration and acceleration capacity, new indicators of autonomic control of the cardiovascular system [47], in patients with heart failure with reduced left ventricular ejection fraction. Patients who were diagnosed as having chronic heart failure with New York Heart Association class I or II, who had a left ventricular ejection fraction ≤40%, and who took an optimal pharmacological therapy for heart failure were included. Twenty-four patients were randomly assigned to participate in the cardiac rehabilitation group and the control group. However, two patients in each group dropped out from the study. A total of 20 patients (10 in the training group and 10 in the control group) completed the study protocol. The mean ages of the training and control groups were 59 ± 7.92 and 56.5 ± 8.43 years, respectively. The mean BMIs of the training and control groups were 27.17 and 28.57 kg/m^2, respectively. Patients in the training group engaged in supervised training three times a week for 24 weeks. The training session consisted of 10 min of warm-up walking, 20 min of breathing exercises and nonresistance arm and leg movements, 20 min of circuit RT using a mechanical bicycle, and 5 min of cool-down stretching. The initial intensity of the bicycle ergometer was 50% of the peak workload in the exercise test, and it was gradually increased up to the maximum workload achieved in the initial exercise test or 80% of maximum heart rate. After the 24-week cardiac rehabilitation, in the training group patients, the mean resting R-R interval was prolonged, the high- and low-frequency band of the power spectral HRV analysis had increased, and the magnitudes of deceleration and acceleration capacity had also increased, whereas the low-frequency/high-frequency index and the standard deviation of the R-R intervals did not change. In the control group patients, no significant changes in HRV-derived measures were observed. The reduction in deceleration and acceleration capacity, which are regulated by the balance between the sympathetic and parasympathetic nervous activity, is a good predictive value for mortality in patients with heart failure [47] and type 1 diabetes [48]. Cardiac rehabilitation, including circuit RT, could improve ANF in patients with chronic heart failure. Sacre et al. [46] investigated the efficacy of exercise training to improve cardiac autonomic function in type 2 diabetic patients with non-ischemic subclinical left ventricular dysfunction. Patients with valvular disease, ischemic heart disease, cardiovascular disease, psychiatric disorder, symptomatic macro- or micro-vascular complications, and low ejection fraction (<50%) were excluded. A total of 49 patients were randomly allocated to the exercise intervention group ($n = 25$) and the control group ($n = 24$). Two subjects were lost to follow-up, and 49 patients were included in the intention-to-treat analysis. The exercise program was based on the recommendation for type 2 diabetes. The patients performed 20–40 min of aerobic exercise and 6–12 RTs twice a week for 24 weeks. The exercise intensity was monitored using rating of perceived exertion and targeted to a moderate to vigorous intensity. Unfortunately, the details of the RT were not described in the paper. The mean ages of patients in the exercise and control groups were 59 ± 10 and 60 ± 9 years, respectively. The mean BMIs of the patients in the exercise and control groups were 32 ± 6 and 32 ± 5 kg/m^2, respectively. The study participants had fair glycemic control; i.e., their hemoglobin A1c levels were approximately 7.7%. Diastolic blood pressure was significantly higher in the exercise group at baseline. Waist circumference was significantly decreased in the exercise group with a significant difference between groups. The VO_{2peak} increased by 11% in the exercise intervention group compared with −1% in the control group. Patients in the exercise group also showed a significant increase in metabolic equivalents (+29%) compared with controls (+2%). The improvement in exercise capacity was not accompanied with a change in HRV (coefficient of variation of normal R-R intervals). Exercise elicited a significant decrease in resting heart rate and an increase in the reciprocal mean R-R interval. The standard deviation of normal R-R intervals and the total spectral power increased in the exercise intervention group, and total spectral power had a significant group effect. However, baroreflex sensitivity did not change in both groups. Ejection fraction and left ventricular mass were also unchanged in both groups. Although the coefficient of variation of the normal R-R interval did not change in this study, other markers of HRV improved without changes in cardiac function, suggesting that the recommended exercise improved cardiac sympathovagal balance in patients with type 2 diabetes. Exercise improves HRV in healthy individuals [42]; specifically, EE decreases sympathetic nerve activity and increases parasympathetic nerve activity, leading to a risk reduction in CVD [10]. This study showed that exercise also improves ANF in patients with type 2 diabetes and that RT probably contributes to it. However, the effect of RT alone on ANF in type 2 diabetic patients has not been elucidated. The autonomic nervous system concomitant with vascular smooth muscle cells and endothelial cells plays a pivotal role in vascular recruitment in skeletal muscle. The impairment of ANF may affect muscle contraction by reducing blood flow in exercising muscle [49]. On the other hand, diabetic neuropathy impairs muscle performance by lowering motor nerve conduction, and microvascular complications induce muscle dysfunction by oxidative stress, inflammation, and decreased blood flow [49]. Therefore, investigating whether RT improves ANF as well as muscle function in patients with type 2 diabetes is quite important for the prevention of diabetic complications. Further studies are strongly needed.

6. Conclusions

Whole-body progressive RT can increase muscle mass, strength, and function in older individuals [5], which may prevent sarcopenia. However, high-intensity RT could induce arterial stiffness or a neuroendocrinological response, reduce arterial baroreceptor sensitivity, and not improve ANF in older individuals. On the other hand, local isometric RT, such as handgrip exercises, may improve ANF (**Figure 1**). Limited evidence is available on the effects of RT on ANF in older patients with comorbidities; however, RT in conjunction with EE has a favorable effect on ANF.

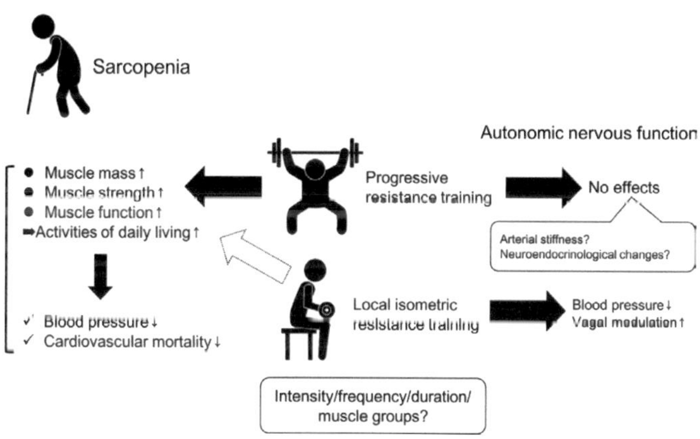

Figure 1. Whole-body progressive resistance training (RT) had no effects on autonomic nervous function (ANF), whereas local isometric RT may improve ANF in older individuals. However, to prevent and/or improve sarcopenia and to reduce the risk of cardiovascular disease and mortality, whole-body progressive RT may be required.

The clinical studies that the author reviewed were well designed; however, the sample sizes were small and many subjects dropped out of the studies (approximately 10–40% of study participants). It is still questionable whether RT is a tolerable intervention for elderly subjects. Moreover, the optimal intensity, frequency, and duration of RT to improve ANF in older individuals remain unknown. The muscle groups that should be trained are also unknown. However, progressive RT with an intensity of approximately 80% of 1RM performed 2–5 times per week for 12–24 weeks seems to have no effect on ANF. Such progressive training may be too hard to improve ANF in older individuals. We should also assess the efficacy of light- (i.e., 10–40% of 1RM) to moderate-intensity (i.e., 40–60% of RM) RT on ANF in the elderly. A systematic review and meta-analysis suggested that RT should be performed twice a week to promote muscular hypertrophy [50]; thus, the frequency should probably be twice a week or more. However, the optimal intensity, frequency, and duration of RT to gain muscle mass and strength may be inconsistent with the RT needed to improve ANF in older individuals.

Clinicians should be prepared for the aging of society. RT can improve activities of daily living as well as muscle strength and function in very elderly people [51]. Older individuals usually suffer from physical disturbances/deconditioning or non-communicable diseases, such as diabetes, hypertension, dyslipidemia, and cardiovascular disease. RT may reduce blood pressure, heart disease, and stroke mortality in subjects with metabolic syndrome [52]. This indicates that RT also improves the prognosis of older individuals with comorbidities. On the other hand, autonomic imbalance (resting heart rate and HRV) is predictors of hypertension, diabetes, the development of cardiovascular disease, and early mortality [53]. If we can determine the parameters of RT to improve ANF, then this would certainly contribute to the health of the elderly. Further studies are warranted in the future.

Author details

Hidetaka Hamasaki

Address all correspondence to: hhamasaki78@gmail.com

1 Department of Internal Medicine, National Center for Global Health and Medicine, Kohnodai Hospital, Chiba, Japan

2 Hamasaki Clinic, Kagoshima, Japan

References

[1] Rosenberg IH. Sarcopenia: origins and clinical relevance. J Nutr. 1997;67:990S–991S.

[2] Janssen I, Heymsfield SB, Ross R. Low relative skeletal muscle mass (sarcopenia) in older persons is associated with functional impairment and physical disability. J Am Geriatr Soc. 2002;50:889–896.

[3] Gariballa S, Alessa A. Sarcopenia: prevalence and prognostic significance in hospitalized patients. Clin Nutr. 2013;32:772–776. doi:10.1016/j.clnu.2013.01.010

[4] Kraschnewski JL, Sciamanna CN, Poger JM, Rovniak LS, Lehman EB, Cooper AB, Ballentine NH, Ciccolo JT. Is strength training associated with mortality benefits? A

15year cohort study of US older adults. Prev Med. 2016;87:121–127. doi:10.1016/j.ypmed.2016.02.038

[5] Liu CJ, Latham NK. Progressive resistance strength training for improving physical function in older adults. Cochrane Database Syst Rev. 2009;(3):CD002759. doi: 10.1002/14651858.CD002759.pub2

[6] Lipsitz LA. Altered blood pressure homeostasis in advanced age: clinical and research implications. J Gerontol. 1989;44:M179–M183.

[7] Esler M, Rumantir M, Kaye D, Jennings G, Hastings J, Socratous F, Lambert G. Sympathetic nerve biology in essential hypertension. Clin Exp Pharmacol Physiol. 2001;28:986–989.

[8] Thomson HL, Wright K, Frenneaux M. Baroreflex sensitivity in patients with vasovagal syncope. Circulation. 1997;95:395–400.

[9] Hart EC, Charkoudian N. Sympathetic neural regulation of blood pressure: influences of sex and aging. Physiology (Bethesda). 2014;29:8–15. doi:10.1152/physiol.00031.2013

[10] Carter JB, Banister EW, Blaber AP. Effect of endurance exercise on autonomic control of heart rate. Sports Med. 2003;33:33–46.

[11] Hu M, Finni T, Zou L, Perhonen M, Sedliak M, Alen M, Cheng S. Effects of strength training on work capacity and parasympathetic heart rate modulation during exercise in physically inactive men. Int J Sports Med. 2009;30:719–724. doi:10.1055/s-0029-1225329

[12] Cooke WH, Carter JR. Strength training does not affect vagal-cardiac control or cardiovagal baroreflex sensitivity in young healthy subjects. Eur J Appl Physiol. 2005;93:719–725. doi:10.1007/s00421-004-1243-x

[13] Heffernan KS, Fahs CA, Shinsako KK, Jae SY, Fernhall B. Heart rate recovery and heart rate complexity following resistance exercise training and detraining in young men. Am J Physiol Heart Circ Physiol. 2007;293:H3180–H3189. doi:10.1152/ajpheart.00648.2007

[14] Taylor AC, McCartney N, Kamath MV, Wiley RL. Isometric training lowers resting blood pressure and modulates autonomic control. Med Sci Sports Exerc. 2003;35:251–256. doi:10.1249/01.MSS.0000048725.15026.B5

[15] Millar PJ, McGowan CL, Cornelissen VA, Araujo CG, Swaine IL. Evidence for the role of isometric exercise training in reducing blood pressure: potential mechanisms and future directions. Sports Med. 2014;44:345–356. doi:10.1007/s40279-013-0118-x

[16] Kingsley JD, Figueroa A. Acute and training effects of resistance exercise on heart rate variability. Clin Physiol Funct Imaging. 2016;36:179–187. doi:10.1111/cpf.12223

[17] Lahiri MK, Kannankeril PJ, Goldberger JJ. Assessment of autonomic function in cardiovascular disease: physiological basis and prognostic implications. J Am Coll Cardiol. 2008;51:1725–1733. doi:10.1016/j.jacc.2008.01.038

[18] Lipsitz LA, Mietus J, Moody GB, Goldberger AL. Spectral characteristics of heart rate variability before and during postural tilt. Relations to aging and risk of syncope. Circulation. 1990;81:1803–1810.

[19] Malik M, Bigger JT, Camm AJ, Kleiger RE, Malliani A, Moss AJ, Schwartz PJ. Heart rate variability. Standards of measurement, physiological interpretation, and clinical use. Task Force of the European Society of Cardiology and the North American Society of Pacing and Electrophysiology. Eur Heart J. 1996;17:354–381.

[20] Sacre JW, Jellis CL, Marwick TH, Coombes JS. Reliability of heart rate variability in patients with type 2 diabetes mellitus. Diabet Med. 2012;29:e33–e40. doi:10.1111/j.1464-5491.2011.03557.x

[21] Tsuji H, Venditti FJ Jr, Manders ES, Evans JC, Larson MG, Feldman CL, Levy D. Reduced heart rate variability and mortality risk in an elderly cohort. The Framingham Heart Study. Circulation. 1994;90:878–883.

[22] Tsuji H, Larson MG, Venditti FJ Jr, Manders ES, Evans JC, Feldman CL, Levy D. Impact of reduced heart rate variability on risk for cardiac events. The Framingham Heart Study. Circulation. 1996;94:2850–2855.

[23] Nolan J, Batin PD, Andrews R, Lindsay SJ, Brooksby P, Mullen M, Baig W, Flapan AD, Cowley A, Prescott RJ, Neilson JM, Fox KA. Prospective study of heart rate variability and mortality in chronic heart failure: results of the United Kingdom heart failure evaluation and assessment of risk trial (UK-heart). Circulation. 1998;98:1510–1516.

[24] Madden KM, Levy WC, Stratton JK. Exercise training and heart rate variability in older adult female subjects. Clin Invest Med. 2006;29:20–28.

[25] Miyachi M. Effects of resistance training on arterial stiffness: a meta-analysis. Br J Sports Med. 2013;47:393–396. doi:10.1136/bjsports-2012-090488

[26] Jensen-Urstad K, Reichard P, Jensen-Urstad M. Decreased heart rate variability in patients with type 1 diabetes mellitus is related to arterial wall stiffness. J Intern Med. 1999;245:57–61.

[27] Melo RC, Quitério RJ, Takahashi AC, Silva E, Martins LE, Catai AM. High eccentric strength training reduces heart rate variability in healthy older men. Br J Sports Med. 2008;42:59–63. doi:10.1136/bjsm.2007.035246

[28] Pratley R, Nicklas B, Rubin M, Miller J, Smith A, Smith M, Hurley B, Goldberg A. Strength training increases resting metabolic rate and norepinephrine levels in healthy 50- to 65-yr-old men. J Appl Physiol. 1994;76:133–137.

[29] Raastad T, Glomsheller T, Bjøro T, Hallén J. Changes in human skeletal muscle contractility and hormone status during 2 weeks of heavy strength training. Eur J Appl Physiol. 2001;84:54–63.

[30] Miyachi M, Kawano H, Sugawara J, Takahashi K, Hayashi K, Yamazaki K, Tabata I, Tanaka H. Unfavorable effects of resistance training on central arterial compliance: a randomized intervention study. Circulation. 2004;110:2858–2863.

[31] Mazzeo RS. Aging, immune function, and exercise: hormonal regulation. Int J Sports Med. 2000;21:S10–S13.

[32] Guarner-Lans V, Rubio-Ruiz ME, Pérez-Torres I, Baños de MacCarthy G. Relation of aging and sex hormones to metabolic syndrome and cardiovascular disease. Exp Gerontol. 2011;46:517–523. doi:10.1016/j.exger.2011.02.007

[33] Aguilera G. HPA axis responsiveness to stress: implications for healthy aging. Exp Gerontol. 2011;46:90–95. doi:10.1016/j.exger.2010.08.023

[34] Takahashi AC, Melo RC, Quitério RJ, Silva E, Catai AM. The effect of eccentric strength training on heart rate and on its variability during isometric exercise in healthy older men. Eur J Appl Physiol. 2009;105:315–323. doi:10.1007/s00421-008-0905-5

[35] Jensen-Urstad K, Storck N, Bouvier F, Ericson M, Lindblad LE, Jensen-Urstad M. Heart rate variability in healthy subjects is related to age and gender. Acta Physiol Scand. 1997;160:235–241.

[36] Karavirta L, Tulppo MP, Laaksonen DE, Nyman K, Laukkanen RT, Kinnunen H, Häkkinen A, Häkkinen K. Heart rate dynamics after combined endurance and strength training in older men. Med Sci Sports Exerc. 2009;41:1436–1443. doi:10.1249/MSS.0b013e3181994a91

[37] Peng CK, Havlin S, Stanley HE, Goldberger AL. Quantification of scaling exponents and crossover phenomena in nonstationary heartbeat time series. Chaos. 1995;5:82–87.

[38] Pikkujämsä SM, Mäkikallio TH, Sourander LB, Räihä IJ, Puukka P, Skyttä J, Peng CK, Goldberger AL, Huikuri HV. Cardiac interbeat interval dynamics from childhood to senescence: comparison of conventional and new measures based on fractals and chaos theory. Circulation. 1999;100:393–399.

[39] Saito M, Iwase S, Hachiya T. Resistance exercise training enhances sympathetic nerve activity during fatigue-inducing isometric handgrip trials. Eur J Appl Physiol. 2009;105:225–234. doi:10.1007/s00421-008-0893-5

[40] Martinez DG, Nicolau JC, Lage RL, Toschi-Dias E, de Matos LD, Alves MJ, Trombetta IC, Dias da Silva VJ, Middlekauff HR, Negrão CE, Rondon MU. Effects of long-term exercise training on autonomic control in myocardial infarction patients. Hypertension. 2011;58:1049–1056.

[41] Kanegusuku H, Queiroz AC, Silva VJ, de Mello MT, Ugrinowitsch C, Forjaz CL. High-intensity progressive resistance training increases strength with no change in cardio-

vascular function and autonomic neural regulation in older adults. J Aging Phys Act. 2015;23:339–345. doi:10.1123/japa.2012-0324

[42] Sandercock GR, Bromley PD, Brodie DA. Effects of exercise on heart rate variability: inferences from meta-analysis. Med Sci Sports Exerc. 2005;37:433–439.

[43] Wanderley FA, Moreira A, Sokhatska O, Palmares C, Moreira P, Sandercock G, Oliveira J, Carvalho J. Differential responses of adiposity, inflammation and autonomic function to aerobic versus resistance training in older adults. Exp Gerontol. 2013;48:326–333. doi: 10.1016/j.exger.2013.01.002

[44] Goodman RA, Posner SF, Huang ES, Parekh AK, Koh HK. Defining and measuring chronic conditions: imperatives for research, policy, program, and practice. Prev Chronic Dis. 2013;10:E66. doi:10.5888/pcd10.120239

[45] Ricca-Mallada R, Migliaro ER, Piskorski J, Guzik P. Exercise training slows down heart rate and improves deceleration and acceleration capacity in patients with heart failure. J Electrocardiol. 2012;45:214–219. doi:10.1016/j.jelectrocard.2012.01.002

[46] Sacre JW, Jellis CL, Jenkins C, Haluska BA, Baumert M, Coombes JS, Marwick TH. A six-month exercise intervention in subclinical diabetic heart disease: effects on exercise capacity, autonomic and myocardial function. Metabolism. 2014;63:1104–1114. doi: 10.1016/j.metabol.2014.05.007

[47] Bauer A, Kantelhardt JW, Barthel P, Schneider R, Mäkikallio T, Ulm K, Hnatkova K, Schömig A, Huikuri H, Bunde A, Malik M, Schmidt G. Deceleration capacity of heart rate as a predictor of mortality after myocardial infarction: cohort study. Lancet. 2006;367:1674–1681. doi:10.1016/S0140-6736(06)68735-7

[48] Guzik P, Piskorski J, Contreras P, Migliaro ER. Asymmetrical properties of heart rate variability in type 1 diabetes. Clin Auton Res. 2010;20:255–257. doi:10.1007/s10286-010-0057-7

[49] Orlando G, Balducci S, Bazzucchi I, Pugliese G, Sacchetti M. Neuromuscular dysfunction in type 2 diabetes: underlying mechanisms and effect of resistance training. Diabetes Metab Res Rev. 2016;32:40–50. doi:10.1002/dmrr.2658

[50] Schoenfeld BJ, Ogborn D, Krieger JW. Effects of resistance training frequency on measures of muscle hypertrophy: a systematic review and meta-analysis. Sports Med. 2016, in press. doi:10.1007/s40279-016-0543-8

[51] Valenzuela T. Efficacy of progressive resistance training interventions in older adults in nursing homes: a systematic review. J Am Med Dir Assoc. 2012;13:418–428. doi: 10.1016/j.jamda.2011.11.001

[52] Lemes ÍR, Ferreira PH, Linares SN, Machado AF, Pastre CM, Netto J Jr. Resistance training reduces systolic blood pressure in metabolic syndrome: a systematic review and meta-analysis of randomised controlled trials. Br J Sports Med. 2016, in press. doi: 10.1136/bjsports-2015-094715

[53] Wulsin LR, Horn PS, Perry JL, Massaro JM, D'Agostino RB. Autonomic imbalance as a predictor of metabolic risks, cardiovascular disease, diabetes, and mortality. J Clin Endocrinol Metab. 2015;100:2443–2448. doi:10.1210/jc.2015-1748

Chapter 7

Exercise Therapy for Acquired Immune Deficiency Syndrome (AIDS) Patients

Jeanne M. Grace

Additional information is available at the end of the chapter

http://dx.doi.org/10.5772/64651

Abstract

Background: Highly active antiretroviral therapy (HAART) has reduced mortality and morbidity in HIV-infected patients and transformed HIV infection from an acute to a chronic disease, not only increasing the life expectancy but also adding to the potential side effects associated with drug therapy. Exercise has been fused as an important adjuvant therapy and can play a valuable role in the management of health in HIV/AIDS patients.

Methods: EBSCOhost, Science Citation Index, CINAHL database, PsycINFO, Cochrane Database of Systematic Reviews, Physiotherapy Evidence Database, and SPORTDiscus were searched between 1996 and May 2016. Searches of published and unpublished abstracts were conducted, as well as a hand search of reference lists and tables of contents of relevant journals and books were conducted. A total of 60 studies met the inclusion criteria.

Results: Most studies failed to indicate the optimum type (mode), intensity, frequency, and duration of aerobic excercise (AE), progressive resistive exercise (PRE), and a combination of AE and PRE (CARE) prescribed to HIV-infected individuals in relation to the different clinical stages of the disease.

Conclusion: The findings provide information that clarifies the optimal mode, duration, frequency, and intensity of AE, PRE, and CARE prescribed to the different clinical stages of HIV patients. The exercises recommended have the potential to benefit HIV patients and should be adopted by clinical exercise therapists.

Keywords: aerobic exercise, CD4 count, highly active antiretroviral therapy, immunodeficiency, progressive resistive exercise

1. Introduction

Medical advances in the treatment of the HIV disease with highly active antiretroviral therapy (HAART) converted HIV infection into a chronic condition that has been related to several comorbidities [1], disability, impairment in daily activities, and decreased exercise capacity [2, 3]. Increased longevity in the HAART era has dramatically reduced mortality and morbidity of HIV-infected patients [4, 5]. Although HAART has dramatically reduced the prevalence of the wasting syndrome and immunosuppression in HIV patients [6], HAART is associated with anthropometric and metabolic conditions including insulin resistance, dyslipidemia, and abnormal distribution of body fat [5, 7, 8].

Exercise is a key management strategy employed by rehabilitation health professionals for health promotion and rehabilitation of patients with HIV/AIDS [9, 10]. Exercise has the potential to ameliorate a range of side effects associated with HIV infection, as well as the cardiometabolic and morphological complications (i.e., mitochondrial dysfunction, inflammation, and oxidative stress) that may accompany HAART. A large body of scientific evidence suggests that exercise can delay the progression of the disease, improve quality of life [11, 12], improve aerobic capacity [3, 9, 13], improve functional ability [14], have beneficial effects on insulin resistance and diabetes [15], improve oxidative stress [13, 16], improve muscle strength [9, 17], improve lipid profile [13, 18], and reduce the risk of cardiovascular disease [1, 19] in people with HIV infection. Of importance is that exercise is generally regarded as safe and there is no evidence to suggest that regular exercise will suppress immune function in asymptomatic and symptomatic individuals with HIV/AIDS [10, 20]. Exercise training, however, apparently confers no beneficial effect on HIV status and viral load [20, 21].

Exercise studies in HIV/AIDS patients indicate that aerobic exercise (AE) [19, 20] and progressive resistive exercise (PRE) [22] improve various health indices in HIV-infected patients with a combination of aerobic and progressive resistive exercise (CARE) more effective than applying only one component [10]. CARE has recently been employed and recommended by the American College of Sports Medicine [23].

2. HIV disease stages and classification

It is of critical importance to recognize HIV disease stages and classification systems not only for tracking and monitoring the HIV epidemic but also for providing clinicians and patients with important information about the clinical management of the disease. Clinical exercise therapists should therefore consider the HIV disease stages as well as the varying needs during each stage in order to adjust the exercise prescription (Ex Rx) accordingly. Two major classification systems currently are in use: The World Health Organization Clinical Staging and Disease Classification System and the US Centres for Disease Control and Prevention (CDC) classification system [24]. The WHO Clinical Staging and Disease Classification System are used in resource-constrained settings without access to CD4 cell count measurements or other diagnostic and laboratory testing methods whereas the CDC disease staging system assesses the severity of HIV disease by CD4 cell counts and by the presence of specific HIV-related conditions. The CDC disease staging system should be used because it incorporates the CD4 cell count categories; knowledge of which is important for the clinical exercise therapist because this may dictate changes in Ex Rx.

The CDC categorization of HIV/AIDS is based on the lowest documented CD4 cell count and on previously diagnosed HIV-related conditions (**Table 1**).

CD4 cell count categories	Clinical categories		
	A Asymptomatic, acute HIV, or PGL	B* Symptomatic conditions, not A or C	C# AIDS-indicator conditions
(1) ≥500 cells/µL	A1	B1	C1
(2) 200–499cells/µL	A2	B2	C2
(3) <200 cells/µL	A3	B3	C3

Abbreviations: PGL = persistent generalized lymphadenopathy

* Category B Symptomatic Conditions - Consult the AIDS Education and Training Centre [24].

Category C AIDS-Indicator Conditions - Consult the AIDS Education and Training Centre [24].

Note: Reprinted with permission from AETC.

Table 1. CDC classification system for HIV-infected adults and adolescents [24].

For example, if a patient had a condition that once met the criteria for category B but now is asymptomatic, the patient would remain in category B. Additionally, categorization is based on specific conditions, as indicated by the AIDS Education and Training Centre [24]. Patients in categories A3, B3, and C1-C3 are considered to have AIDS [24]. The AIDS Education and Training Centre (AETC) can be referred to for a detailed account of the symptomatic conditions for the different clinical categories [24].

3. Pre-exercise evaluation/testing

HIV/AIDS individuals should be screened for the presence, signs, symptoms, and/or risk factors suggestive of cardiovascular, pulmonary, or metabolic diseases, as well as other conditions (e.g., orthopedic limitations, pregnancy) that require special attention to (i) aid in the development of a safe and effective Ex Rx and (ii) optimize safety during exercise testing [23]. Regardless of stage of disease progression, all HIV-infected individuals should consult with their physician and obtain medical clearance prior to participating in an exercise program. Once cleared to exercise, it is recommended that the patient consult a clinical exercise therapist for advice and guidelines on an exercise program tailored to the individual's physical function, health status, exercise response, and stated goals.

	Time point/Frequency								
Lab test	**Entry into Care**	**ART Initiation or Modification**	**2 to 8 Weeks Every 6 Months**	**After ART Initiation or Modification**	**Every 3 to 6 Months**	**Every 12 Months**	**Treatment Failure**	**Clinically Indicated**	**If ART Initiation is Delayed[b]**
CD4 Count	√	√		√ During first 2 years of ART or if viremia develops while patient on ART or CD4 count <300 cells/mm^3		√ After 2 years on ART with consistently suppressed viral load: CD4 Count 300–500 cells/mm^3: •Every 12 months CD4 Count >500 cells/mm^3: •CD4 monitoring is optional	√	√	Every 3-6 month
Fasting lipid profile[c]	√	√			√ If abnormal at last measurement	√ If abnormal at last measurement		√	√ If normal at baseline, annually
Fasting Glucose or Hemoglobin A1C	√	√		√ If abnormal at last measurement		√ If abnormal at last measurement		√	√ If normal at baseline, annually

[a] If ART initiation occurs soon after HIV diagnosis and entry into care, repeat baseline laboratory testing is not necessary.

[b] ART is indicated for all HIV-infected individuals and should be started as soon as possible. However, if ART initiation is delayed, patients should be retained in care, with periodic monitoring as noted above.

[c] Consult the National Lipid Association's recommendations for management of patients with dyslipidaemia

Table 2. Laboratory monitoring schedule for HIV-infected patients before and after initiation of antiretroviral therapy[a] [37].

A comprehensive pre-exercise test evaluation for HIV-infected individuals includes (i) medical history, (ii) physical examination, and (iii) HIV-related laboratory tests. The physician should provide medical information (e.g., stage of disease, CD4 count, HAART, and other medications used, history of symptoms, and recent illness), whereas the clinical exercise therapist can acquire exercise-related medical information by using the Canadian Physician Clearance Form (ePARmed-X+) [25] as a positive response is expected when using the Canadian Activity Readiness Questionnaire for Everyone (PAR-Q+) [26]. The PARmed-X+ is a form developed for use by physicians and clinical exercise therapists to assist in addressing medical concerns regarding physical activity participation that were identified by the PAR-Q+.

Health-related quality of life (HRQOL) has emerged as an important topic in the treatment of HIV infection, with HIV-infected patients generally demonstrating a HRQOL that is lower than that of the general population [11, 27]. Clinical exercise therapists should use the Multidimensional Quality of Life Questionnaire for Persons with HIV (MQOLHIV) to monitor HRQOL for patients living with HIV disease [28]. HRQOL has become increasingly important with the goals of therapy now, including improvement of HRQOL, in addition to reduction of symptoms, virus suppression, and extension of survival. Repeated assessment can be used to track changes in functional status over time; monitor, assess, and optimize treatment effects and enhance communication between patient and provider with the potential to improve the health care process and overall survival [28]. With physical training regarded as a non-pharmacological treatment, use of HRQOL will be helpful to monitor and improve adherence [29] because the development of side effects can worsen HRQOL and lead to treatment non-adherence [30].

Regardless of the disease status of HIV-infected individuals, it is important to assess their current fitness level prior to developing an exercise program [31]. Despite the lower functional capacities of HIV-infected individuals [11], the standard physical fitness tests applicable to the apparently healthy populations can be applied [31]. BP and ECG are monitored during exercise testing due to the increased prevalence of cardiovascular impairments and cardiac dysfunction [23].

Components of the physical examination conducted by the clinical exercise therapist should include the following:

1. Body composition: It includes body weight, body mass index (BMI), skin fold measurement of subcutaneous fat (fat percentage), standard circumference sites (thigh, calf, arm, chest, waist, and hip), waist-to-hip ratio (WHR) [21, 23], muscle mass, lean body mass [32], and self-reported body shape changes [33]. Perceived body image can be measured with the Body Image Scale (BIS) that is a 12-item, disease-specific, valid and reliable tool that measures perceived body image along five dimensions (comfort, competence, appearance, predictability, and existential self). Each item utilizes a five point visual analogue scale and scores range from 12 to 60 (higher scores reflect poorer body image) [34].

2. Physical capacity: The modified Bruce protocol or submaximal YMCA cycle ergometer test is used to determine cardiorespiratory fitness [23]. For assessing muscular strength, six-repetition maximum (6 RM) or 10 RM protocol may be more appropriate because the

affected population is generally untrained [31]. A grip strength dynamometer can be used to determine the maximum isometric strength of the hand and forearm muscles [17].

3. Neuromuscular function: Gait analysis and balance test (e.g., Stork stand) should be conducted because peripheral neuropathy, which can be a consequence of fast hyperlactatemia at rest which many people living with HIV/AIDS present, is a common symptom of HIV infection [31, 35].

4. Functional capacity: The Lower Extremity Functional Scale (LEFS) can be used as a measure of patients' initial function, ongoing progress, and outcome, as well as used to set functional goals [36].

3.1. Laboratory tests

To assist the clinical exercise therapist, it is recommended that a physician conduct the following laboratory tests for HIV-infected individuals:

1. Cardiometabolic: It is fasting blood lipids (total cholesterol, LDL-C, HDL-C, and triglycerides), fasting glucose or hemoglobin A1c (HbA1C), and CD4 cell count when diagnosed with HIV as well as prior to the start of HAART [10, 21]. **Table 2** indicates the laboratory monitoring schedule for HIV-infected patients before and after initiation of antiretroviral therapy[a] [37].

2. Bone mineral density (BMD): The literature recommends that for all HIV-infected postmenopausal women and men aged > 50 years, BMD should be tested and, if the results of the test do not warrant medical treatment, the test should be repeated every 2–5 years, depending on the proximity to thresholds for therapy [38].

3. Electrocardiography: Due to the increased risk of heart disease, a resting 12-lead electrocardiogram is recommended [31].

4. Special considerations prior to prescribing an exercise program

Clinical exercise therapists should consider various factors (e.g., clinical/treatment characteristics) when prescribing an exercise program for HIV-infected individuals. HIV-infected individuals adapt readily to exercise training and therefore can the general FITT principle of Ex Rx for apparently healthy adults applied to those carrying HIV [23, 39]. Additional drug-related physical, psychological, and physiological side effects including lethargy, nausea, vomiting, diarrhea, headaches, fever, muscle pain, occasional dizziness, peripheral neuropathy, fatigue, anemia, mitochondrial toxicity, depression, and myopathy must be considered [37]. In particular, the following symptoms, adverse effects, and comorbidity should be taken into account:

4.1. HIV/AIDS wasting syndrome

It is a syndrome of unintentional weight loss that occurs with advanced HIV infection. It is defined as an involuntary weight loss of greater than 10% from baseline body weight during the previous 12 months or a 5% loss of weight during the previous 6 months [40].

4.2. Lipodystrophy

It is as an all-encompassing term used to describe a metabolic complication of fat loss, fat gain, or a combination of fat loss and gain, which is associated with some antiretroviral (ARV) therapies given to HIV-infected individuals [41]. *Lipoatrophy* is the loss of subcutaneous fat, particularly evident in the face, buttocks, and limbs. *Lipohypertrophy* is the accumulation of visceral and central fat in the abdomen, dorsocervical region ("buffalo hump"), or breasts. *Fat redistribution* is an all-encompassing term used to describe lipohypertrophy, lipoatrophy, and/or mixed syndrome in HIV-infected individuals [42, 43].

4.3. Dyslipidemia

HAART is associated with hypercholesterolemia, increases in LDL-C, hypertriglyceridemia, and lowered HDL-C [2, 44].

4.4. Diabetes

HAART adversely affects insulin sensitivity and glucose tolerance [45, 46]. Clinical exercise therapists should also distinguish between the three subgroups of patients with diabetes: those with pre-existing diabetes who contract HIV; those who have diabetes at onset of HIV infection; and others who develop hyperglycemia after commencing HAART. These subgroups need to be managed differently. Pre-existing type 2 diabetes mellitus may continue to be managed, after diagnosis of HIV, with the same drug therapy that was being used prior to detection of HIV [46]. It is important to counsel these patients about a possible deterioration in metabolic function, and the chances of drug interactions between oral antidiabetic drugs and HAART. Patients diagnosed with diabetes and HIV together may be treated according to guidelines for uninfected individuals. Patients developing diabetes after HAART may benefit from insulin, because it is a safe and effective method of treatment [46].

4.5. BMD

Initiation of ART is associated with a 2–6% decrease in BMD over the first 2 years [47]. HIV-infected individuals receiving protease-inhibitor-based HAART (up to 30%) are more likely to display significant bone demineralization [48].

4.6. Cardiometabolic disease (CMD) risk

The incidence of CMD increases in HIV-infected individuals, and HAART is associated with an increase in both peripheral and coronary artery disease [49, 50]. Risk factors such as hyperlipidemia [44], oxidative stress [51], impaired glucose tolerance, and increased insulin resistance [45], accumulation of visceral fat [52], inflammation secondary to HIV [53], and the effects of some antiretroviral drugs all contribute to the risk of developing CMD [54].

5. Program prescription

Exercise training is well established as an important non-pharmacological therapy for HIV-infected individuals and one of the key approached being explored to deal with the complications and symptoms of HIV/AIDS [9, 55]. The primary goals (depending on the stage of the disease) for prescribing exercise to HIV-infected patients are to improve HRQOL, physical, and physiological, functional, and neuromuscular function capacity, decrease the risk of CMD, and promote long-term exercise compliance. It is therefore advised to include both short- and long-term individualized goals based on the subjective and objective findings in the patient's assessment. In addition to the symptoms, adverse effects, and comorbidity considerations mentioned previously, clinical exercise therapists should take into account the individual's: (i) physical function; (ii) functional limitations and likes/dislikes; (iii) health status; (iv) availability of equipment and time available to train; (v) exercise dose response (desired goal, type of exercise, and intensity, duration, and frequency of training); and (vi) coordination among members of the multidisciplinary team [10, 23]. HIV/AIDS is a progressive disease; therefore, clinical exercise therapists should frequently reassess the patient's physical, physiological, functional, and neuromuscular capacity (at least every 6–8 weeks) to determine whether the exercise program is still meeting the individual needs and goals of the patient. As mentioned earlier, HIV-infected individuals adapt readily to exercise training, and therefore, the general FITT principle of Ex Rx for apparently healthy adults can be applied to those carrying HIV [23, 39]. An Ex Rx is the process whereby the recommended exercise regimen is designed in a systematic and individualized manner in terms of the Frequency (How Often?), Intensity (How Hard?), Time (How Long?), and Type/Mode (What Kind?), or the FITT principle of Ex Rx [56].

5.1. Frequency

Frequency refers to how often an individual engages in an activity, usually the number of days per week [56]. Most HIV studies have used a three times weekly intervention for both the aerobic exercise (AE) and progressive resistive exercise (PRE) component [12, 13, 57] with success three/four times per week [31], a few studies have reported positive results twice weekly [39, 58]. **Table 3** indicates the specific frequency of the AE and PRE component according to the different clinical categories with the corresponding CD4 cell count for each category [55]; for example, HIV-infected individuals in clinical category A1 (asymptomatic, acute HIV, or persistent generalized lymphadenopathy with ≥500 cells/μL) (**Table 1**) and clinical category A2 (asymptomatic, acute HIV, or persistent generalized lymphadenopathy with 200–499 cells/μL) (**Table 1**) can perform PRE three times per week and aerobic AE five times per week whereas clinical category C HIV-infected individuals (AIDS indicator conditions with <200 cells/μL) (**Table 1**) can perform PRE twice per week and AE three times per week (**Table 3**) [55]. There should be a rest day between the PRE sessions and, if time constraints are present, separate resistive programs into upper and lower body workouts.

http://dx.doi.org/10.5772/64651

It is recommended to complete flexibility exercises during the warm-up and cool-down of an exercise session. Neuromuscular exercises can be performed two to three times per week by those HIV-infected individuals who are severely debilitated from the disease, typically those from the Centres for disease control clinical categories A3, B3, and C1-C3 (**Table 1**).

5.2. Intensity

Intensity is the level of exertion experienced during the activity. Individuals can monitor their rate of perceived exertion (RPE) with Borg's RPE scale that subjectively rates exercise intensity from 6 to 20 with 6 corresponding to "no exertion at all" and 20 corresponding to "maximal exertion" [59]. A person can also use a definitive scale by tracking heart rate. For the aerobic component, a 40–60% exercise intensity of VO_2R (difference between VO_{2max} and resting VO_2) or heart rate reserve (HRR) [31] or 50–85% of HR_{max} [10] is recommended to HIV-positive individuals.

The recommended exercise intensity for the PRE component is 50–90% of one-repetition maximum (1RM) [12]; 60% of 1RM [23] with an intensity of 35–50% of 6–10RM for patients from the Centres for disease control clinical categories A3, B3, and C1–C3 (**Table 3**) [55]. The specific intensities of the PRE component according to the different clinical categories with the corresponding CD4 cell count for each category are displayed in (**Table 3**) [55].

Clinical Category	Physical Fitness classifications[a]	Frequency[b]		Intensity[c]			Time		Type (mode)
		Kcal. wk^{-1}	d. wk^{-1}	HRR/$\dot{V}O_2R$	% HR_{max}	Perception of effort[d]	Total Weekly Duration per day (min)	Total Weekly Duration per week (min)	
A (Asymptomatic, acute HIV, or PGL)	Average–good	>2,000	5 **(AE)** 3 **(PRE)**	65%–80% **(AE)**; 65%–85% of 1RM or 2–3 sets of 10–15 reps **(PRE)**	80%–91% **(AE)**	Moderate–hard	30–90 **(AE + PRE)**	200–300	**AE**: Cycle, walk, jog, row **PRE**: Target major muscle groups using weights, cables/pulleys/resistance bands, bodyweight and plyometrics
B (Symptomatic conditions not A or C)	Fair–average	1,500–2,000	3–5 **(AE)** 2–3 **(PRE)**	55%–70% **(AE)**; 55%–85% of 1RM or 2 sets of 10–12 reps **(PRE)**	74%–84% **(AE)**	Moderate	30–90 **(AE + PRE)**	200–300	**AE**: Cycle, walk, row **PRE**: Target major muscle groups using weights, cables/pulleys/resistance bands, bodyweight
C (AIDS – Indicator conditions)	Poor	500–1,000	3 **(AE)** 2 **(PRE)**	30%–45% **(AE)**; 35%–50% of (6–10RM) or 1–2 sets of 8–10 reps **(PRE)**	57%–67% **(AE)**	Light–moderate	20–30 **(AE + PRE)**	60–150	**AE**: Cycle, walk, hydrotherapy **PRE**: Target major muscle groups using cables/pulleys/resistance bands, bodyweight

Kcal–kilocalories; $\dot{V}O_2R$–oxygen uptake reserve; HRR–heart rate reserve; $\%HR_{max}$–% age–predicted maximal heart rate; AE–aerobic exercise; PRE–progressive resistive exercise; PGL–persistent generalized lymphadenopathy.

[a] Fitness classification based on normative fitness data categorized by $\dot{V}O_{2max}$.

[b] Maximum frequency (d.wk^{-1}) of combined AE and PRE for clinical category A = 5; B = 5; C= 3.

[c] The various methods to quantify exercise intensity in this table may not necessarily be equivalent to each other.

[d] Perception of effort using the rating of perceive exertion (RPE) [23] or talk test [23].

Include 5 to 10 minutes of pre-exercise warm-up and cool-down exercises.

Table 3. Recommended FITT framework for the frequency, intensity, time, and type of aerobic exercise (AE) and progressive resistive exercise (PRE) for different HIV clinical categories [55].

5.3. Time

Time refers to the duration of an exercise session [56] and depends on the intensity as well as the goal of the exercise. AE performed at a higher intensity performed for a shorter duration produces the same results as exercise done at a moderate intensity for a longer duration. A duration of 45 min for the AE component [12] or 30–60 min [23] including a 5–10 min warm-up and cool-down is recommended. The duration of the AE component for patients from the Centres for disease control clinical categories A3, B3, and C1-C3 is 20–30 min (**Table 3**) [55].

The recommended duration of the PRE component is 30–90 min with a volume of three sets of eight reps [12], two sets of eight to ten repetitions [31], or 30 min to complete two to three sets of 10–12 exercises [23]. One to two sets of eight to ten repetitions are recommended for patients from the Centres for disease control clinical categories A3, B3, and C1–C3 (**Table 3**) [55].

The duration of the flexibility component is 5–10 min with a volume of three sets of 20–30 s. Similarly, the duration of the neuromuscular component is 5–10 min with the volume determined by the specific neuromuscular exercise.

5.4. Types (modes) of exercise

Types of activity include aerobic, balance, flexibility, and resistance (or strength) training which help patients achieve different health goals [56]. In HIV-infected individuals, the appropriate exercise mode depends on the patient's functional impairments, preference, and safety issues regarding the stage of the disease [12, 60].

5.4.1. AE

Aerobic exercise is safe to perform (even at high intensities) and may lead to significant improvements in selected outcomes of cardiovascular fitness, body composition, and psychological status in HIV-infected individuals [19]. Individual studies have shown an improvement in VO_{2max} and other measures of fitness [10, 13], improvement in body composition (decreases in fat percentage, increases in leg muscle area and abdominal fat loss), [13, 21] reduced anxiety, stress, and depressive mood [22], improvement in QOL [11], as well as improvement in lipid profile [8] and glucose tolerance [21]. Crossover effects have also been reported whereby aerobic exercise may lead to improved strength [3, 19]. Type or mode of AE to improve the cardiovascular system includes running, walking, swimming, aerobics classes, circuit training, and cycling [55].

5.4.2. PRE

Progressive resistive exercise is safe and effective in medically stable adults living with HIV [22, 32]. PRE can increase body weight and peripheral girth, and reverse muscular atrophy [61]. Training effects include an increase in strength [17, 61], improvements in body composition (increasing BMI and lean body mass) [17, 62], and improvements in psychological status and

QOL [10], insulin sensitivity [63], and lipid profile [64]. PRE that stresses the muscular system includes weights, cables/pulleys/resistance bands, body weight, and plyometrics [31, 55].

5.4.3. Combined aerobic and resistive exercise (CARE)

The combination of the two exercise modalities, AE and PRE, is recommended by the American College of Sports Medicine [23] and seems not only to be safe but well tolerated by HIV-infected individuals. A combination of aerobic and progressive resistive exercise (CARE) appears to be more effective than applying only one component [10]. Research studies have shown an improvement in VO_{2max} and other measures of fitness [3, 10, 13], functional capacity [10, 14], improvement in body composition (decreases in fat percentage, increases in abdominal fat loss, BMI, and lean body mass) [8, 13, 57], increases in viral load and cell counts for TCD4+/TCD8 [13], improvement in QOL [8, 9, 11] and mood state [61], as well as improvement in lipid profile [13] and glucose tolerance [21].

6. Special considerations during exercise training

For asymptomatic HIV-infected individuals, the ACSM's guidelines regarding contraindications for exercise applicable to healthy individuals should apply as there are no current guidelines for individuals with HIV/AIDS [23].

It is important that HIV-infected individuals report increased feelings of tiredness or exhaustion (minor increases in feelings of fatigue should not be a reason for exercise discontinuation), lower gastrointestinal distress (especially diarrhea), shortness of breath, or a sense of increased effort in performing activities of daily living [31].

Dizziness, swollen joints, or vomiting preclude exercise participation [23].

The exercise sessions of symptomatic individuals with HIV/AIDS or those diagnosed with comorbidities should be supervised [23].

7. Conclusion

Evidence-based research indicates that AE, PRE, and a combination thereof by HIV-infected individuals are safe and positively influences the cardiometabolic and morphological complications that may accompany HAART as well as a range of side effects associated with the HIV disease itself. The current exercise guidelines available in the literature are generalized and they do not consider the specific clinical stages with significant gaps in knowledge as to the minimal and optimal duration, frequency, mode, and intensity of exercise needed to produce beneficial changes in HIV patients. Most of the studies referred to have not approached the exercise program from a dose-response perspective or from the perspective of validating commonly used exercise prescriptions for HIV-infected individuals. The research provides information that clarifies the optimal mode, duration, frequency, and intensity of AE and PRE prescribed to the different clinical stages of HIV patients. Clinical exercise therapists should adopt the recommended guidelines that will benefit HIV patients.

Author details

Jeanne M. Grace

Address all correspondence to: Gracej@ukzn.ac.za

Discipline of Biokinetics, Exercise and Leisure Sciences, College of Health Sciences, University of Kwazulu Natal, Westville, Durban, South Africa

References

[1] Neto MG, Zwirtes R, Brites C. A literature review on cardiovascular risk in human immunodeficiency virus-infected patients: implications for clinical management. The Brazilian Journal of Infectious Diseases. 2013;17: 691–700. doi:10.1016/j.bjid.2013.05.004

[2] Giannarelli C, Klein RS, Badimon JJ. Cardiovascular implications of HIV-induced dyslipidemia. Atherosclerosis. 2011;219: 384–389. doi:10.1016/j.atherosclerosis.2011.06.003

[3] O'Brien KK, Tynan AM, Nixon SA, Glazier RH. Effectiveness of aerobic exercise for adults living with HIV: systematic review and meta-analysis using the Cochrane Collaboration protocol. BMC Infectious Diseases, 2016;16: 182. doi:10.1186/s12879-016-1478-2

[4] Peters BS, Conway K. Therapy for HIV past, present, and future. Advances in Dental Research. 2011;23: 23–27. doi:10.1177/0022034511399082

[5] Blanco F, San Román J, Vispo E, Lopez M, Salto A, Abad V, Soriano V. Management of metabolic complications and cardiovascular risk in HIV-infected patients. AIDS Rev. 2010;2: 231–241. PMID: 21179187

[6] Anuurad E, Semrad A, Berglund L. Human immunodeficiency virus and highly active antiretroviral therapy-associated metabolic disorders and risk factors for cardiovascular disease. Metabolic Syndrome and Related Disorders. 2009;7: 401–409. doi:10.1089/met.2008.0096

[7] Garg A. Lipodystrophies: genetic and acquired body fat disorders. The Journal of Clinical Endocrinology & Metabolism. 2011;96: 3313–3325. doi:10.1210/jc.2011-1159#sthash.i21n4Idx.dpuf

[8] Ogalha C, Luz E, Sampaio E, Souza R, Zarife A, Neto MG, Netto E, Brites C. A randomized, clinical trial to evaluate the impact of regular physical activity on the quality of life, body morphology and metabolic parameters of patients with AIDS in Salvador, Brazil. JAIDS Journal of Acquired Immune Deficiency Syndromes. 2011;57: S179–S185. doi:10.1097/QAI.0b013e31821e9bca

[9] Gomes-Neto MG, Conceição CS, Carvalho VO, Brites C. Effects of combined aerobic and resistance exercise on exercise capacity, muscle strength and quality of life in HIV-infected patients: a systematic review and meta-analysis. PloS One. 2015;10: e0138066. doi:10.1371/journal.pone.0138066

[10] Gomes-Neto M, Conceicao CS, Carvalho VO, Brites C. A systematic review of the effects of different types of therapeutic exercise on physiologic and functional measurements in patients with HIV/AIDS. Clinics. 2013;68: 1157–1167. doi:10.6061/clinics/2013(08)16

[11] Mbada CE, Onayemi O, Ogunmoyole Y, Johnson OE, Akosile CO. Health-related quality of life and physical functioning in people living with HIV/AIDS: a case–control design. Health and Quality of Life Outcomes. 2013;11: 1–8. doi: 10.1186/1477-7525-11-106

[12] Gomes Neto M, Ogalha C, Andrade AM, Brites C. A systematic review of effects of concurrent strength and endurance training on the health-related quality of life and cardiopulmonary status in patients with HIV/AIDS. BioMed Research International. 2013;3: 1–8. doi:10.1155/2013/319524

[13] Garcia A, Fraga GA, Vieira Jr RC, Silva CM, Trombeta JC, Navalta JW, Prestes J, Voltarelli FA. Effects of combined exercise training on immunological, physical and biochemical parameters in individuals with HIV/AIDS. Journal of Sports Sciences. 2014;32: 785–792. doi:10.1080/02640414.2013.858177

[14] Mendes EL, Andaki AC, Amorim PR, Natali AJ, Brito CJ, Paula SO. Physical training for HIV positive individuals submitted to HAART: effects on anthropometric and functional parameters. Revista Brasileira de Medicina do Esporte. 2013;19: 16–21. doi: 10.1590/S1517-86922013000100003

[15] Bevilacqua M, Dominguez LJ, Barbagallo M. Insulin Resistance and the cardiometabolic syndrome in HIV infection. Journal of The Cardiometabolic Syndrome. 2009;4: 40–43. doi:10.1111/j.1559-4572.2008.00027.x

[16] Deresz LF, Lazzarotto AR, Manfroi WC, Gaya A, Sprinz E, Oliveira ÁR, Dall'Ago P. Oxidative stress and physical exercise in HIV positive individuals. Revista Brasileira de Medicina do Esporte. 2007;13: 275–279. doi:10.1590/S1517-86922007000400013

[17] Kietrys D, Galantino ML. Can progressive resistive exercise improve weight, limb girth, and strength of individuals with HIV disease? Physical Therapy. 2014;94: 329–333. 10.2522/ptj.20120466

[18] Stradling C, Chen YF, Russell T, Connock M, Thomas GN, Taheri S. The effects of dietary intervention on HIV dyslipidaemia: a systematic review and meta-analysis. PloS One. 2012;7: e38121. doi:10.1371/journal.pone.0038121

[19] O'Brien K, Nixon S, Tynan AM, Glazier R. Aerobic exercise interventions for adults living with HIV/AIDS. Cochrane Database Systematic Review. 2010;8: 1–7. doi: 10.1002/14651858.CD001796.pub3

[20] Hand GA, Lyerly GW, Jaggers JR, Dudgeon WD. Impact of aerobic and resistance exercise on the health of HIV-infected persons. American Journal of Lifestyle Medicine. 2009;3: 489–499. doi:10.1177/1559827609342198

[21] Fillipas S, Cherry CL, Cicuttini F, Smirneos L, Holland AE. The effects of exercise training on metabolic and morphological outcomes for people living with HIV: a systematic review of randomised controlled trials. HIV clinical trials. 2010;11: 270–282. doi:10.1310/hct1105-270

[22] O'Brien K, Tynan AM, Nixon S, Glazier RH. Effects of progressive resistive exercise in adults living with HIV/AIDS: systematic review and meta-analysis of randomized trials. AIDS care. 2008;20: 631–653. doi:10.1080/09540120701661708

[23] Thompson PD, Arena R, Riebe D, Pescatello LS. American College of Sports Medicine. ACSM's guidelines for exercise testing and prescription. 9th Ed. Philadelphia: Lippincott Williams & Wilkins; 2013. 445 p. doi:10.1249/JSR.0b013e31829a68cf

[24] AETC. AIDS Education and Training Centre Guide for HIV/AIDS Clinical Care. [Internet]. 2014. Available from: http://aidsetc.org/guide/hiv-classification-cdc-and-who-staging-system [Accessed 2016-04-21]

[25] ePARmed-X+. ePARmed-X+ Physician Clearance Follow-Up. [Internet]. 2015. Available from: http://eparmedx.com/wp-content/uploads/2013/03/ePARmedX-Physician-Clearance-Form-2014.pdf [Accessed 2016-04-21]

[26] The Physical Activity Readiness Questionnaire for Everyone. [Internet]. 2015. Available from: http://eparmedx.com/wp-content/uploads/2013/03/PARQplusJan2015.pdf [Accessed 2016-04-20]

[27] Imam MH, Flora MS, Moni MA, Shameem RK, Haque MA, Mamun SA. Health related Quality of Life with HIV/AIDS in different stages of HIV infection. Mymensingh medical journal: MMJ. 2012;21: 509–515. PMID: 22828552

[28] Clayson DJ, Wild DJ, Quarterman P, Duprat-Lomon I, Kubin M, Coons SJ. A comparative review of health-related quality-of-life measures for use in HIV/AIDS clinical trials. Pharmacoeconomics. 2006;24: 751–765. PMID: 16898846

[29] Burgoyne RW, Tan DH. Prolongation and quality of life for HIV-infected adults treated with highly active antiretroviral therapy (HAART): a balancing act. Journal of antimicrobial chemotherapy. 2008;61: 469–473. doi:10.1093/jac/dkm499

[30] Gordon LL, Gharibian D, Chong K, Chun H. Comparison of HIV virologic failure rates between patients with variable adherence to three antiretroviral regimen types. AIDS patient care and STDs. 2015;29: 384–388. doi:10.1089/apc.2014.0165

[31] Durstine JL. American college of sports medicine's exercise management for persons with chronic diseases and disabilities. 3rd Ed.Champaign: Human Kinetics; 2009. 439 p. doi:10:0736074333

[32] Sakkas GK, Mulligan K, DaSilva M, Doyle JW, Khatami H, Schleich T, Kent-Braun JA, Schambelan M. Creatine fails to augment the benefits from resistance training in patients with HIV infection: a randomized, double-blind, placebo-controlled study. PLoS One. 2009;4: e4605. doi:10.1371/journal.pone.0004605

[33] Hsue PY, Squires K, Bolger AF, Capili B, Mensah GA, Temesgen Z, Wanke CA, Wohl DA. Screening and assessment of coronary heart disease in HIV-infected patients. Circulation. 2008;118: e41–47. doi:10.1161/CIRCULATIONAHA.107.189626

[34] Law M, Friis - Møller N, Weber R, Reiss P, Thiebaut R, Kirk O, D'Arminio Monforte A, Pradier C, Morfeldt L, Calvo G, El - Sadr W. Modelling the 3-year risk of myocardial infarction among participants in the Data Collection on Adverse Events of Anti - HIV Drugs (DAD) study. HIV medicine. 2003;4: 1–10. doi:10.1046/j.1468-1293.2003.00138.x

[35] Prentice WE, Kaminski TW. Rehabilitation techniques for sports medicine and athletic training. 5th Ed. New York, NY: McGraw-hill; 2004. 850 p. doi:10.0073376612

[36] Binkley JM, Stratford PW, Lott SA, Riddle DL. The Lower Extremity Functional Scale (LEFS): scale development, measurement properties, and clinical application. Physical Therapy. 1999;79: 371–383. PMID: 10201543

[37] Aidsinfo. HIV/AIDS drug side effects [Internet] 2016. Available from: https://aidsinfo.nih.gov/contentfiles/lvguidelines/adultandadolescentgl.pdf [Accessed 2016-04-27]

[38] Aberg JA, Kaplan JE, Libman H, Emmanuel P, Anderson JR, Stone VE, Oleske JM, Currier JS, Gallant JE. Primary care guidelines for the management of persons infected with human immunodeficiency virus: 2009 update by the HIV medicine Association of the Infectious Diseases Society of America. Clinical infectious Diseases. 2009;49: 651–681. doi:10.1086/605292

[39] Hand GA, Phillips KD, Dudgeon WD, William Lyerly G, Larry Durstine J, Burgess SE. Moderate intensity exercise training reverses functional aerobic impairment in HIV-infected individuals. AIDS Care. 2008;20: 1066–1074. doi:10.1080/09540120701796900

[40] Selik RM, Mokotoff ED, Branson B, Owen SM, Whitmore S, Hall HI. Revised surveillance case definition for HIV infection—United States, 2014. MMWR Recommendations and Reports. 2014;63: 1–10. ISSN: 1057-5987

[41] Finkelstein JL, Gala P, Rochford R, Glesby MJ, Mehta S. HIV/AIDS and lipodystrophy: Implications for clinical management in resource-limited settings. Journal of the International AIDS Society. 2015;18: 19033. doi:10.7448/IAS.18.1.19033

[42] Mercier S, Gueye NF, Cournil A, Fontbonne A, Copin N, Ndiaye I, Dupuy AM, Cames C, Sow PS, Ndoye I, Delaporte E. Lipodystrophy and metabolic disorders in HIV-1-infected adults on 4-to 9-year antiretroviral therapy in Senegal: a case-control study. Journal of Acquired Immune Deficiency Syndromes. 2009;51: 224–230. doi:10.1097/qai.0b013e31819c16f4

[43] van Griensven J, De Naeyer L, Mushi T, Ubarijoro S, Gashumba D, Gazille C, Zachariah R. High prevalence of lipoatrophy among patients on stavudine-containing first-line antiretroviral therapy regimens in Rwanda. Transactions of the Royal Society of Tropical Medicine and Hygiene. 2007;101: 793–798. doi:10.1016/j.trstmh.2007.02.020

[44] Husain NE, Ahmed MH. Managing dyslipidemia in HIV/AIDS patients: challenges and solutions. HIV/AIDS (Auckland, NZ). 2015;7: 1–10. doi:10.2147/HIV.S46028

[45] Gianotti N, Visco F, Galli L, Barda B, Piatti P, Salpietro S, Bigoloni A, Vinci C, Nozza S, Gallotta G, Lazzarin A. Detecting impaired glucose tolerance or type 2 diabetes mellitus by means of an oral glucose tolerance test in HIV-infected patients. HIV Medicine. 2011;12: 109–117. doi:10.1111/j.1468-1293.2010.00860.x

[46] Kalra S, Kalra B, Agrawal N, Unnikrishnan AG. Understanding diabetes in patients with HIV/AIDS. Diabetology & Metabolic Syndrome. 2011;3: 1–7. doi: 10.1186/1758-5996-3-2

[47] McComsey GA, Tebas P, Shane E, Yin MT, Overton ET, Huang JS, Aldrovandi GM, Cardoso SW, Santana JL, Brown TT. Bone disease in HIV infection: a practical review and recommendations for HIV care providers. Clinical Infectious Diseases. 2010;51: 937–946. doi:10.1086/656412

[48] Harris VW, Brown TT. Bone loss in the HIV-infected patient: evidence, clinical implications, and treatment strategies. Journal of Infectious Diseases. 2012;205: S391-S398. doi:10.1093/infdis/jis199

[49] Zhou DT, Kodogo V, Chokuona KF, Gomo E, Oektedalen O, Stray-Pedersen B. Dyslipidemia and cardiovascular disease risk profiles of patients attending an HIV treatment clinic in Harare, Zimbabwe. HIV/AIDS (Auckland, NZ). 2015;7: 145–155. doi:10.2147/HIV.S78523.

[50] Capili B, Anastasi JK, Ogedegbe O. HIV and general cardiovascular risk. Journal of the Association of Nurses in AIDS Care. 2011;22: 362–375. doi:10.1016/j.jana.2010.12.002

[51] Masiá M, Padilla S, Fernández M, Rodríguez C, Moreno A, Oteo JA, Antela A, Moreno S, del Amo J, Gutiérrez F, CoRIS B. Oxidative stress predicts all-cause mortality in HIV-infected patients. PloS One. 2016;11: e0153456. doi:10.1371/journal.pone.0153456

[52] Snopková S, Matýsková M, Povolná K, Polák P, Husa P. HIV lipodystrophy. Vnitrni Lekarstvi. 2010;56: 1217–1222. PMID: 1261108

[53] Kip AE, Balasegaram M, Beijnen JH, Schellens JH, de Vries PJ, Dorlo TP. Systematic review of biomarkers to monitor therapeutic response in leishmaniasis. Antimicrobial agents and chemotherapy. 2015;59: 1–44. doi:10.1128/AAC.04298-14

[54] Sued O, Figueroa MI, Cahn P. Clinical challenges in HIV/AIDS: Hints for advancing prevention and patient management strategies. Advanced Drug Delivery Reviews. 2016;103: 5-19. doi:10.1016/j.addr.2016.04.016

[55] Grace JM, Semple SJ, Combrink S. Exercise therapy for human immunodeficiency virus/AIDS patients: Guidelines for clinical exercise therapists. Journal of Exercise Science & Fitness. 2015;13: 49–56. doi:10.1016/j.jesf.2014.10.003

[56] Oberg E. Physical activity prescription: our best medicine. Integrative Medicine. 2007;6: 18–22.

[57] Ley C, Leach L, Barrio MR, Bassett S. Effects of an exercise programme with people living with HIV: research in a disadvantaged setting. African Journal of AIDS Research. 2014;13: 313–319. doi:10.2989/16085906.2014.961937

[58] Fillipas S, Oldmeadow LB, Bailey MJ, Cherry CL. A six-month, supervised, aerobic and resistance exercise program improves self-efficacy in people with human immunodeficiency virus: a randomised controlled trial. Australian Journal of Physiotherapy. 2006;52: 185–910. doi:10.1016/S0004-9514(06)70027-7

[59] Borg G. Borg's perceived exertion and pain scales. Campaign: Human Kinetics; 1998. 104 p. Accession Number: 1998-07179-000

[60] Baar K. Training for endurance and strength: lessons from cell signaling. Medicine and Science in Sports and Exercise. 2006;38: 1939–1944. doi: 10.1249/01.mss0000233799.62153.19

[61] Shephard RJ. Physical impairment in HIV infections and AIDS: responses to resistance and aerobic training. The Journal of Sports Medicine and Physical Fitness. 2015;55: 1013–1028. PMID: 24947921

[62] Keithley JK, Swanson B. HIV-associated wasting. Journal of The Association of Nurses in AIDS Care. 2013;24: S103-S111. doi:10.1016/j.jana.2012.06.013

[63] Broholm C, Mathur N, Hvid T, Grøndahl TS, Frøsig C, Pedersen BK, Lindegaard B. Insulin signaling in skeletal muscle of HIV-infected patients in response to endurance and strength training. Physiological Reports. 2013;1: e00060. doi:10.1002/phy2.60.

[64] Zanetti HR, Cruz LG, Lourenço CL, Ribeiro GC, Leite MA, Neves FF, Silva-Vergara ML, Mendes EL. Nonlinear resistance training enhances the lipid profile and reduces inflammation marker in people living with HIV: a randomized clinical trial. Journal of Physical Activity and Health. 2016;2: 1543–5474. PMID: 26901242